Morphological and Functional Aspects of Placental Dysfunction

Contributions to Gynecology and Obstetrics

Vol. 9

Series Editor
P.J. Keller, Zürich

S. Karger · Basel · München · Paris · London · New York · Sydney

Morphological and Functional Aspects of Placental Dysfunction

Volume Editor
H. Soma, Tokyo

60 figures and 21 tables, 1982

S. Karger · Basel · München · Paris · London · New York · Sydney

Contributions to Gynecology and Obstetrics

Vol. 6: Real-Time Ultrasound in Perinatal Medicine. *Chef, R.*, Charleroi (ed.)
   VIII + 160 p., 112 fig., 18 tab., 1979. ISBN 3–8055–2976–7
Vol. 7: *Jaluvka, V.*, Berlin. Surgical Geriatric Gynecology
   X + 174 p., 66 tab., 1980. ISBN 3–8055–3070–6
Vol. 8: *Gjorgov, A.N.*, Philadelphia, Pa. Barrier Contraception and Breast Cancer
   X + 162 p., 12 fig., 66 tab., 1980. ISBN 3–8055–0330–X

National Library of Medicine, Cataloging in Publication
   Morphological and functional aspects of placental dysfunction/
   volume editor, H. Soma. – Basel; New York: Karger, 1982.
   (Contributions to gynecology and obstetrics; v. 9)
   1. Maternal-Fetal Exchange 2. Placenta Diseases 3. Placenta Diseases
   – physiopathology
   I. Soma, H. II. Series
   W1 C0778RG v. 9 [WQ 212 M871]
   ISBN 3–8055–3510–4

Drug Dosage
   The author and publisher have exerted every effort to ensure that drug selection and dosage set forth in this text are in accord with current recommendations and practice at the time of publication. However, in view of ongoing research, changes in government regulations, and the constant flow of information relating to drug therapy and drug reactions, the reader is urged to check the package insert for each drug for any change in indications and dosage and for added warnings and precautions. This is particularly important when the recommended agent is a new and/or infrequently employed drug.

All rights reserved
   No part of this publication may be translated into other languages, reproduced or utilized in any form or by any means, electronic or mechanical, including photocopying, recording, microcopying, or by any information storage and retrieval system, without permission in writing from the publisher.

© Copyright 1982 by S. Karger AG, P.O. Box, CH-4009 Basel (Switzerland)
   Printed in Switzerland by Tanner & Bosshardt AG, Basel
   ISBN 3–8055–3510–4

Contents

Introduction VII

Baldwin, V.J. (Vancouver): Morphologic Pathology of Fetomaternal
  Interaction 1
Teasdale, F. (Montreal): Morphometric Evaluation 17
Fisher, C.C. (Paddington): Ultrasonic Placental Ageing 29
Gille, J. (Würzburg): Immunopathological Alterations in the Dysfunctioning Placenta 41
Soma, H.; Yoshida, K.; Mukaida, T.; Tabuchi, Y. (Tokyo): Morphologic Changes in the Hypertensive Placenta 58
Asmussen, I. (Copenhagen): Vascular Morphology in Diabetic Placentas 76
De Wolf, F.; Brosens, I. (Leuven); Robertson, W.B. (London): Ultrastructure of Uteroplacental Arteries 86
Garcia, A.G.P. (Rio de Janeiro): Placental Morphology of Low-Birth-Weight Infants Born at Term: Gross and Microscopic Study of 50 Cases 100
Altshuler, G. (Oklahoma City, Okla.): Placentitis 113
Kullander, S.; Maršál, K.; Persson, P.-H. (Malmö): Human Placental Lactogen and Ultrasonic Screening for the Detection of Placental Insufficiency 129
Nakayama, T.; Yanaihara, T. (Tokyo): Placental Sulfatase Deficiency: Biochemical and Clinical Aspects 145
Lee, J.N.; Chard, T. (London): New Placental Proteins in Placental Dysfunction 157

Subject Index 170

Introduction

As perinatology advances, much knowledge on the human placental function is required. Until recently, obstetricians have shown little interest in placental morphology because this organ was discarded. Indeed, it may seem that such morphological studies have little to contribute to placental function. However, in many ways, the placenta reflects the course of prenatal life and from its structural alterations one may deduce prenatal deleterious influences on the fetus. As the perplexing diversity of placental morphology and function has not been sufficiently explored to make meaningful correlations with abnormal fetal growth, the meaning of placental dysfunction, which might be indicative of a deficiency or impairment of transfer capacity from the mother to the fetus, should be reevaluated. Due to progress in the application of ultrasonography and biochemical investigations, interest has lately become focussed on fundamental and practical knowledge of the interrelationship between the placental morphology and function. But many problems of placental dysfunction still have to be cleared up. This volume collects recent research work on placental morphology and function, and I have been fortunate in obtaining the collaboration of prominent authorities in these fields.

The main topics of this volume are related to the morphologic and ultrastructural bases of placental dysfunction, to the ultrasonic and biochemical evaluation of placental dysfunction and especially to placental infection.

Hopefully, this volume will be helpful in understanding the basis and practical processes of placental dysfunction.

Hiroaki Soma

Morphologic Pathology of Fetomaternal Interaction

Virginia J. Baldwin

Departments of Pathology of the University of British Columbia and
British Columbia Children's Hospital, Vancouver, Canada

Introduction

The morphologic pathology of the placenta differs from that of any other organ because this tissue is a composite of two individuals who, although related, are distinct. Thus, abnormalities identified in the placenta must be interpreted in the context of fetomaternal interaction. Problems in one organism can affect the other, often with specific placental manifestations. Occasionally, primarily placental lesions have secondary effects on the mother and/or the fetus. The following discussion identifies these areas and the pathologic lesions encountered.

The areas to be considered are those maternal conditions which affect the fetus, those fetal conditions which affect the mother, those conditions where there is combined disease or interaction, and placental disorders with both fetal and maternal consequences.

Maternal Conditions

The maternal conditions affecting the fetoplacental unit may be grouped as structural or functional problems of the reproductive tract, genetic disorders, exogenous influences, and maternal disease states.

Structural or Functional Abnormalities of the Reproductive Tract

Significant structural or functional problems in the reproductive tract are deformities of the uterus or pelvis, neuromuscular disorders, incompetence of the uterine cervix and ectopic implantation of the gestation.

Some of these conditions result in delivery of an infant before it is structurally or physiologically mature enough to cope adequately with extrauterine existence. The respiratory distress syndrome is compounded by physiologic immaturity in temperature and metabolic regulation, cardiorespiratory control mechanisms, and renal and gastrointestinal function. The pathology in the preterm infant who develops this cascade of problems can range from minimal hyaline membrane disease in an infant who dies soon after delivery to a constellation of lesions such as bronchopulmonary dysplasia, hepatic lesions of intravenous alimentation, hemorrhage (particularly intracranial) and consequences of sepsis.

The placenta in these cases is small, thin, coarsely textured and pale with appropriate histologic immaturity of the chorionic villi. If preterm labor and delivery have been initiated or complicated by infection or premature separation of the placenta due to decidual necrosis and hemorrhage, these lesions will also be seen.

Others of these maternal conditions interfere with the normal course of labor and delivery, and asphyxial insults to the fetus may be identified by specific associated lesions. The details depend on the intensity of the insult, the duration of the fetal distress and the ability of the fetus to respond. In the fetus, there is involution of the thymus, visceral congestion with petechiae and focal hemorrhages, loss of stainable fat in the fetal adrenal cortex and variably impaired endochondral ossification. With increased fetal activity and respiratory movements there is aspiration of amniotic debris. If meconium has been passed there may be staining of the fetal skin and aspiration of meconium along with the amniotic debris.

The placental manifestations of fetal distress are primarily those of meconium staining and residual evidence of the lesion causing the distress. Varying with the duration and intensity of meconium exposure there may be greenish brown staining of the amnion and degenerative changes of the epithelium. The macrophages of the amnion and chorion phagocytose the meconium and can be identified by their content of orange brown granules. Although meconium staining as evidence of fetal distress is becoming of some medicolegal importance, the correlation of duration of exposure to meconium and extent and intensity of the pathologic findings is not yet exact.

If premature separation has been the origin of the distress, there may be pathologic evidence of this in the form of a focus of decidual necrosis with hemorrhage and possibly associated placental villous infarction, usually at the margin of the disc. An early intrapartum separation may not be identifiable pathologically. A severe asphyxial insult may cause intrauterine fetal death, and with retention of the dead fetus, there is progressive bland obliteration of the fetal vasculature of the villous tissue.

Of the structural maternal problems, the septate or subseptate uterus may be associated with preterm delivery of an otherwise normally progressing gestation, due to volume constraints of the divided uterine cavity. Other structural lesions of the uterus such as large uterine leiomyomata, particularly near the cervix, may interfere with delivery and lead to fetal asphyxia. In both these situations, placental development may be anomalous in relation to the uterine lesion with unusual shapes such as extra lobes or fenestrae. Also, the decidual development may be impaired over the uterine anomaly causing the placenta to be unusually adherent at that site with muscle fibers identified in the decidua of the delivered placenta.

Congenital or acquired deformities of the maternal pelvis or neuromuscular diseases could impede normal labor and delivery, leading to asphyxial damage unless the situation is recognized and an appropriate delivery route chosen.

The incompetent uterine cervix is associated with the threat of preterm delivery often accompanied by ascending infection of the amniotic cavity. If the labor does not occur for some time after the cervix begins to open there may be rupture of membranes with chronic leakage of amniotic fluid. This results in oligohydramnios, and the consequent constrained intrauterine volume has deformational effects on skeletal, facial and pulmonary development of the fetus [29], recognized as the Potter phenotype, originally described with oligohydramnios due to renal agenesis. With oligohydramnios the placenta may also contain amnion nodosum but it is usually less extensive than in oligohydramnios due to anuria. This lesion consists of areas of denudation of amnion with deposition of mounds of amniotic debris, variably covered by proliferating amnion from the margins of these mounds.

Ectopic gestations may be ectopic in site or depth of implantation. The majority of extrauterine ectopics are in the oviduct and occur on a clinical background suggesting structural or functional impairment of tubal transport of the developing ovum [16]. This site is inappropriate in size and lacks decidua so the invading trophoblast rapidly penetrates the wall with disruption and hemorrhage. Detailed morphologic and cytologic examination of the available products of these gestations often demonstrates an abnormality of the embryo as well, such as incomplete embryonic development, chromosomal error or structural malformation.

Intrauterine ectopics are usually low implantations and of unknown pathogenesis. The placentas are broad and thin due to less adequate decidual development in the lower uterine segment, and the portion near or over the cervix is prone to premature separation. This is seen in the delivered placenta as an area of decidual necrosis with old or recent hemorrhage, often with inflammation. If there has been external bleeding before delivery there is associated ascending infection manifest by inflammation on the placental surface.

Abnormal depth of implantation due to over-penetration of the uterine wall complicates delivery of the placenta rather than the fetus, but rarely the entire wall is penetrated and this may be associated with antenatal hemorrhage and fetal asphyxia. When hysterectomy is required, the adherence of chorionic villi to myometrium with only intervening fibrin is demonstrable pathologically. Focal and minor degrees of adherence are relatively common and muscle fibers may be seen in what must be a hypoplastic area of decidua on the delivered placenta, often without a clinical history of difficulty with delivery of the placenta.

Related to this is the undetermined influence of the uterine decidua on the development of the placenta in general. There are a variety of placental shapes that are encountered and the exact mechanism of their formation is not known, although intrauterine masses or septa may be suggested and the bilobed placenta may represent a sulcus implant. These variations are significant when they are associated with complications, such as bleeding compounded by ascending infection. This is particularly striking with the severe forms of extrachorial attachment of the placental membranes [4], the circumvallate placenta, in which there may also be reduced fetal vascularization of the chorionic surface.

Genetic Factors

The genetic influence of the maternal organism arises in several areas. The increased risk of chromosomal aneuploidy or polyploidy in the fertilized oocyte of the older mother is sufficiently great that antenatal screening with amniocentesis is now available in many communities to women over 35 years of age.

The influence of maternal sex-linked gene disorders on the fetus will depend on which maternal X chromosome the fetus receives and, if female, whether the maternal or paternal X chromosome is inactivated and in what tissues.

With improved detection and medical care, the mother with serious genetic disease is surviving to a reproductive age. The best studied example is the mother with phenylketonuria. These women are having offspring with intrauterine and postnatal growth retardation, mental retardation and congenital malformations due to the abnormal metabolic milieu occasioned by the maternal metabolic error [26].

Couples with chronically unsuccessful reproductive histories are being identified as having balanced translocation in one or other of the parents. This is leading to unbalanced chromosomal abnormalities in the fetus, manifest especially as an increased incidence of spontaneous early gestational wastage [30].

Finally, any heritable disorder of genetic or chromosomal origin in the maternal oocyte can be passed on to the fetus with the manifestations depending on the pattern of inheritance and the paternal contribution.

The fetal manifestations of maternal genetic disease will vary with the specific situation. With the use of antenatal diagnostic techniques, pathologists are being called on to confirm diagnoses in material obtained at elective termination of pregnancy, and sophisticated biochemical, cytogenetic and morphologic techniques are being applied to these fetuses, as the morphologic manifestations of some of the common genetic errors are much more subtle and less specific in early gestation [2].

The placental manifestations of maternal genetic disease depend on the disorder under consideration. It has been suggested that there are specific morphologic findings associated with polyploidy and aneuploidy, particularly dysmorphogenesis and depressed growth [14]. Metabolic errors in the fetus, usually recessively inherited from the parents, are occasionally identifiable in the placenta with vacuolation of trophoblast or stromal cells due to stored material [25]. Nonimmunologic fetoplacental hydrops has been associated with a variety of fetal malformations and disorders, some of which have genetic origins, such as 45,X gonadal dysgenesis and Meckle's syndrome.

Exogenous Influences

The whole discipline of teratology is concerned with the effect of exogenous influences on gestation, and by virtue of the location of the gestation, these influences are necessarily exerted through the mother.

Although the fetus is in a relatively protected position when considering external trauma, blunt trauma may exert severe shearing forces that may cause separation of the placenta. Depending on the speed and severity of the separation there will be evidence of fetal asphyxia to the point of intrauterine death. Direct penetrating injury will have manifestations depending on the route, the mechanism and other specific features.

With the development of antenatal diagnostic and therapeutic maneuvers, an important potential source of iatrogenic exogenous influences has emerged. In a competent and experienced center, with appropriate localization techniques, the hazards of amniocentesis, direct fetal visualization techniques and intrauterine therapy such as transfusions can be minimized [6]. However, complications such as fetal or placental hemorrhage, introduction of infection, direct fetal trauma or stimulation of preterm labor and delivery may occur with attendant lesions of the fetus and placenta.

The other large group of exogenous influences that originate with the mother and her environment are those that form the basis of much of teratologic research and include irradiation, abnormal temperatures and an ever expanding number of chemicals. The delineation of the dose- and time-specific anomalies seen with the phocomelia of thalidomide encouraged the concept of time- and organ-specific teratogenic susceptibility. However, the more thorough epidemiologic and experimental analyses that have been stimulated by the thalidomide association have shown that the teratogenic influence of most agents is likely quite diffuse and not time-restricted, with early pregnancy loss and nonspecific intrauterine growth retardation associated with most. There is likely an element of individual susceptibility to the effect of an otherwise harmless compound and long-term teratogenesis such as transplacental carcinogenesis is now being considered. Closer attention to the maternal environment during gestation and the identification and analysis of clusters of anomalies are required, as it is becoming more difficult to identify specific teratogens with certainty.

While the offspring may manifest various anomalies associated with a specific teratogen, such as nasal deformities and epiphyseal stippling due to warfarin, or vaginal adenosis and carcinoma from diethylstilbestrol exposure, placental findings are not usually reported. It must not be inferred from this, however, that there are no placental lesions with intrauterine exposure to exogenous compounds. Experimental studies on animals have suggested impaired placental development and degenerative changes in association with diethylstilbestrol [28], and heroin has also been recorded to reduce placental growth [11]. A recent survey of placentas from pregnancies associated with alcohol abuse suggested an increase in minor anomalies, vascular lesions, meconium staining and inflammation [3]. Detailed descriptions of the placenta are not yet available for most agents, likely due in part to the small numbers of cases. A general increase in nonspecific findings may be seen in placentas from heavy smokers and users of 'street' drugs, while steroids, hormones, heparin, antiepileptic and antineoplastic therapy is not yet identifiable with a specific pattern of findings.

Maternal Systemic Disease

If the mother is severely ill, reproductive capacity is one of the first functions to be affected, but there are some maternal disorders that may allow, but affect, pregnancy.

Of the maternal vascular diseases, hypertension of pregnancy and/or preexisting benign essential or renovascular hypertension are important. The mechanism of the potential fetal problems is impairment of the maternal vascular supply to the placenta with consequent effects on placental function and hence fetal growth [13].

Mild hypertension may have no identifiable effect but with increasing severity, intrauterine growth retardation may occur. This is manifest as late gestational growth failure, as the affected placenta is less and less able to nourish the fetus. This is differentiated, therefore, from intrauterine growth impairment due to fetal constitutional abnormalities such as chromosomal errors, which is manifest throughout gestation. The nutritionally deprived fetus is small for gestational age in weight more than length. Head size is appropriate for gestational age, as brain growth is usually maintained at the expense of other organs. These infants do not withstand the stresses of labor and delivery well so there may be manifestations of asphyxia. The intrauterine deprivation may be severe enough to cause intrauterine death.

The placental morphology of these gestations is quite variable and tends to be most pronounced when toxemia of pregnancy is superimposed on prepregnancy hypertension. The chronic intervillous hypoxia due to impaired maternal circulation through the placenta creates a dysmature villous pattern consisting of accentuation of syncytial knot formation, focal degeneration of trophoblast with fibrin deposits, and accelerated development of small, apparently mature, but poorly vascularized terminal villi. Intervillous circulation may be focally interrupted with consequent villous infarction. Occasional villous infarcts may be seen in otherwise normal placentas but they are greater in number with severely hypertensive pregnancies. When maternal arterioles are identified in the decidua of the delivered placenta, there may be endovascular proliferative changes and atherosis in cases with severe hypertension [5]. Recent electron-microscopic studies have confirmed that the placental pathology of maternal hypertension is due to uteroplacental ischemia [17].

Maternal anemia has been associated with complications of pregnancy but the mechanism is not understood. Folic acid deficiency has been suggested as important regarding premature separation and hence fetal asphyxia and death, and nonspecific maternal anemia has been associated with nonimmunologic fetoplacental hydrops. With maternal sickle cell trait or disease, the placenta may contain more than the usual amount of maternal intervillous blood, retained because of sickling in the relatively hypoxic intervillous space. Thrombotic thrombocytopenic purpura, which has a predilection for the gestational state, is associated with consumptive coagulopathy, anemia, bleeding, neuropsychiatric disturbances and multiple microinfarcts in maternal tissues. Intrauterine death may occur but there are no specific lesions of the infant or placenta. Thalassemia minor and other disorders of platelets, such as idiopathic thrombocytopenic purpura or antiplatelet antibodies, are not associated with specific lesions.

Maternal malignancy during pregnancy is rare. Leukemia may be identified in the maternal blood in the intervillous space and rarely a solid maternal neoplasm such as melanoma metastasizes to the placenta. Aside from possibly a higher rate of single umbilical artery cords in mothers with mitral stenosis, maternal heart disease has no specific lesions of the placenta. No significant findings have been made in placentas from women with renal disease without hypertension, including solitary kidney and renal transplantation. Crohn's disease, ulcerative colitis and hepatitis have not had unusual manifestations in the placenta. In cholestasis of pregnancy, alterations of trophoblast and reduced intervillous space have been seen [9]. These cases also have more meconium staining but this may be due to the higher rate of fetal distress. Various neuromuscular, cutaneous and pulmonary conditions have been seen but in numbers too few to draw conclusions.

Of the maternal metabolic disorders affecting gestation, diabetes mellitus is the most important, but in nearly half of such pregnancies there is no detectable effect on fetoplacental growth, development or viability [19].

In a third of cases, the macrosomic but physiologically immature infant is encountered with the placenta similarly affected. This is seen more with the less severe stages of maternal diabetes and seems to progress most rapidly after 30 weeks gestation. The fetal hypersomatism includes hypertrophy as well as hyperplasia and the placenta is large, bulky and histologically immature for dates with a proliferative villous pattern. There may also be bland chorionic vessel thrombosis in the diabetic placenta.

In the 12% of cases that die suddenly and unexpectedly in utero at 34–36 weeks of gestation there is usually little if any identifiable abnormality aside from evidence of intrauterine asphyxia with no cause found. This situation may be associated with a diabetic mother with more severe vascular disease.

The diabetic pregnancy is associated with an increased risk of fetal anomalies, up to 22% in women with elevated hemoglobin A_{1C} in the first trimester [21]. A number of anomalies have been described with specially increased risk of cardiovascular, neural tube and lower limb defects. Placental anomalies in diabetic pregnancies are single umbilical artery and hypervascularity of the chorionic villi (chorangiosis).

Finally, the diabetic with severe vascular disease may compromise the gestation in a manner analogous to the hypertensive woman.

Other maternal endocrine abnormalities do not appear to be associated with specific fetoplacental lesions as too few cases of Cushing's disease, Addison's disease, adrenogenital syndrome, hyperparathyroidism, hyperprolactinemia and other rare disorders have been seen to draw conclusions.

Maternal 'connective tissue' disorders are not usually associated with specific fetal or placental abnormalities nor is their steroid therapy. A striking exception is the intrauterine fetal cardiac arrhythmia due to conduction abnormalities through the atrioventricular node in infants of mothers with systemic lupus erythematosus in pregnancy. Although the morphologic pathology of the fetus is not specific, decidual vascular lesions of a necrotizing or inflammatory nature have been described [1]. This suggests that when vasculopathy is an important part of the collagen disorder, nonspecific decidual vascular lesions and evidence of impaired intervillous maternal blood flow similar to that seen with hypertension can be anticipated.

Fetal Conditions

It may seem unusual to consider that fetal difficulties could influence the placenta or the mother but there are several instructive specific instances.

Multiple Gestation

With multiple gestation, the presence of two or more fetuses in the uterus can be associated with an increased incidence of toxemia and obstetric problems related to labor and delivery, with prematurity and asphyxia occurring more frequently than with single pregnancies.

If the gestation is monozygous in origin with two fetuses in one chorionic sac, there may be sufficient imbalance of blood flow through the anastomoses of the chorionic circulations that a twin-to-twin transfusion occurs. Depending on the volume of shunting there may be death of one twin during gestation, birth of living but ill plethoric and pale twin pairs, or apparently unaffected twins. Fluctuations in maternal blood flow may also influence twin survival. Occasionally, acute fetofetal transfusion manifests as hydramnios in the second trimester [33].

When there is early death of a twin embryo, it often remains identifiable in the membranes as a plaque of involuted placenta and flattened macerated but identifiable embryonic tissue. When identifiable, the membranes are usually monochorionic diamniotic, and the presence of such a fetus papyraceus has been associated with skin defects in the surviving twin [20]. When intrauterine death is later in gestation, it may be possible to identify sites of vascular anastomoses although injection studies are unlikely. Secondary effects on the surviving twin have also been described in this situation with apparently discordant disruptive structural defects [15]. At birth of surviving twins from a monochorial gestational sac, the chorionic circulations may be seen to anastomose to a variable extent. Direct surface anastomoses are less significant than intraparenchymal shunts and the most useful areas to study for villous anastomoses are those fields where an artery of one twin approaches the vein of another on the placental surface.

In a special form of the twin-to-twin transfusion syndrome where one twin has a two-vessel cord, the various bizarre acardiacus amorphus fetuses may be seen, as the twin with three vessels in the cord takes over and reverses the circulation in the other, with consequent regressive developmental anomalies [18].

Anomalies

While many fetal anomalies are clinically silent during pregnancy, others are associated with intragestational manifestations and some with specific placental lesions.

Any impairment of fetal urine excretion leads to oligohydramnios, reduced intrauterine volume and secondary compressive effects on the fetus. The presence of amnion nodosum in the placenta in these cases is a clue to the pathogenesis of the respiratory distress these infants manifest due to the associated pulmonary hypoplasia.

Obstructive lesions of the fetal gastrointestinal tract impair resorption of amniotic fluid, and hydramnios is encountered. This of itself is not associated with any specific placental lesion.

Hydramnios accompanied by fetoplacental hydrops has been associated with a wide variety of conditions [12]. As erythroblastosis fetalis due to Rh incompatibility is declining, nonimmunologic causes of fetoplacental hydrops are becoming important. Intrauterine heart failure may be associated with abnormal hemoglobins, fetomaternal transfusion, twin-to-twin transfusion syndrome, cardiac tumors, fetal arrhythmias, arteriovenous shunts such as massive teratomas, or interruptions of the fetal circulatory pathway in the heart. The placenta in these cases is bulky, pale, edematous and contains immature edematous proliferative chorionic villi. Where the cause is associated with destruction of fetal red cells or an abnormal hemoglobin there may be pronounced intracapillary erythroblastosis. No additional specific pathologic finding is described in the other conditions. Particularly disappointing is the lack of other diagnostic placental abnormality in fetomaternal transfusion. Detection of fetal red cells in the maternal circulation is essential in these cases.

Nonimmune hydrops due to impairment of fetal vascular flow in the fetus may be seen with intrathoracic masses such as diaphragmatic hernias or cystic adenomatoid malformations of the lung, intra-abdominal obstruction of the portal circulation with thrombosis, congenital hepatic fibrosis, or compression by masses such as cystic kidneys or neuroblastoma. Additional placental findings in specific cases may be neuroblastomatous metastases in the villi or amnion nodosum with cystic kidneys. Fluid retention may be the mechanism of fetoplacental hydrops with congenital nephrosis, lymphangiectasis of Turner's syndrome but no mechanism has been identified in congenital infections, other chromosomal errors, storage diseases, open central nervous system defects, osteochondrodystrophies or maternal disorders such as toxemia, anemia or diabetes mellitus. The possible additional placental findings in these disorders include evidence of inflammation such as villitis, stored material, or changes already described with maternal disorders and chromosomal errors. There still remains a large group of infants, however, who are hydropic with bulky hydropic immature placentas, in whom no cause is identified.

As indicated, congenital tumors of the fetus may be accompanied by placental findings in the form of hydrops or evident villous metastases. Congenital leukemia is identifiable in the villous circulation and requires careful differentiation from the leukoerythroblastosis due to other causes such as fetal red cell hemolysis [32].

Combined Conditions

There are two important complications of gestation where both mother and fetus contribute to the placental findings – immunologic incompatibility and infection.

Immunologic Incompatibility

Fetomaternal incompatibility of red cell surface antigens is a source of fetoplacental pathology but has now been rendered preventable. The occasional case is still seen, however, and the infant may be edematous and erythroblastotic or undergrown and icteric. The placenta is also edematous with villi that are still proliferating and appear immature for dates. The degree of intracapillary erythroblastosis is variable but seems least when the fetus is most severely affected. Occasionally, there is evidence of complications of diagnostic and therapeutic maneuvers.

Infection

This area of fetomaternal interaction has been well reviewed [7] and consists of two main patterns of disease: ascending infection of the amniotic cavity and hematogenous contamination of villous tissue.

The ascending infection is usually associated with premature rupture of membranes, but may occur through intact membranes, often with an incompetent cervix. The agents are the normal genital flora, often organisms of relatively low virulence and are usually bacterial although viral, mycotic, or mycoplasmal organisms may be implicated. Both the fetus and the mother respond to ascending infection by an acute inflammatory response chemotactically directed toward the amniotic cavity. With severe inflammation, there is opacification of the amnion and fetal vessels while milder inflammation may be marked by edema of the amnion only, particularly in mature placentas. The inflammation consists of variable acute fetal vasculitis in the umbilical cord and on the chorionic surface and penetration of the chorion and amnion by maternal acute inflammatory cells from vessels in the membranes or from the intervillous space. The intensity and patterns of the inflammation are variable with only occasional agent-specific appearances. A special pattern of chronic intrauterine inflammation is occasionally seen with a rim of calcifying debris external to the fetal vessels in the cord, with lesser degrees of chronic inflammation in the membranes and on the placental surface. This lesion is associated with a higher incidence of perinatal morbidity and mortality [22] although the pathogenetic sequence is not yet clarified. Ascending infection is to be differentiated from infection acquired by the fetus during passage through the birth canal. The latter is often due to a nonhemolytic streptococcus or herpes virus and the placenta in these situations is usually free of inflammation.

Hematogenous colonization of chorionic villi may stimulate an inflammatory lesion called villitis. The majority of these agents are thought to be viral although bacteria, protozoa and fungi may be important. There may be no gross placental changes or there may be nonspecific changes such as intervillous thrombi, infarcts and villous pallor. Occasionally, a widespread lesion may manifest as coarsening of the villous texture or thickening of the maternal floor. Microscopically, the lesion is a placentitis or villitis which involves individual villi or clusters of villi. There is necrosis of trophoblast with fibrin deposition, infiltration of the stroma with inflammatory cells, activation of Hofbauer cells and degenerative changes of fetal vessels. There may be associated intravillous hemorrhage and calcification. The acute exudative, proliferative and destructive phase may progress to a chronic picture with villous sclerosis and atrophy. Adjacent affected villi may be agglutinated by fibrin into a large mass. The causative agent is usually not identifiable and the inflammatory picture is in the majority nonspecific. In those instances where evidence of specific agents is seen such as with rubella, cytomegalovirus or toxoplasma, there is some correlation with related disease in the fetus. Where more nonspecific lesions are present, correlations are more difficult, particularly as the number and size of the placental lesions do not always correlate with severity of fetal or maternal illness. This aspect of fetomaternal placental inflammation has been examined in detail in a recent review [27].

Placental Conditions

While the placenta is affected by the many disorders already referred to, it is as a reaction to the fetal or maternal condition. There are, however, a few situations where the primary abnormality is in the placenta.

Tumors of the placenta are uncommon. Choriocarcinoma and hydatidiform mole have been recorded associated with a fetoplacental unit [8, 10] and for the most part influence the maternal condition by virtue of their neoplastic and endocrine characteristics. Chorangiomas of the placenta, if large, act as arteriovenous shunts and may cause fetoplacental hydrops and intrauterine death [31].

Aneurysms of the chorionic vessels are also uncommon but they occasionally rupture causing intrauterine asphyxia and death.

Abnormalities in the development of the body stalk may be manifest in the placenta. Absence of one umbilical artery is not necessarily accompanied by other placental lesions but may be an indication of other fetal anomalies. Complications associated with unusually long or short cords often manifest as intrauterine asphyxia. The absence of a true umbilical cord may be significant in the pathogenesis of the accompanying severe fetal abnormalities [24].

If the amnion ruptures in utero, free strands of amniochorial tissues may entwine fetal parts leading to the constrictions or amputations of the amniotic band syndrome or disruption complex. In other patterns of this lesion there may be adherence of deficient areas of the anterior body wall or cranial vault to the surface of the placenta. There tends to be oligohydramnios associated with these gestations leading to additional deformational fetal anomalies. Pathologically, the fetal surface of the placenta demonstrates a lack of the normal sheen due to amnion. This is born out histologically by a complete loss of the layers of amnion with a tendency to sclerosis of the remaining chorion. There may be an associated inflammatory lesion of the placental surface. Occasionally, small strands of residual amnion can be identified attached to the placental surface. Also occasionally, the amnion will rupture and remain as a small sac around the insertion of the umbilical cord without any apparent deleterious affects on the fetus. The identification of deficiencies of the amnion in these cases is critical to accurate genetic counselling as this is a sporadic intrauterine deformational disorder without known hereditary significance or definite teratogenic origin [23].

Summary

The thorough and interested pathologic examination of the placenta may explain events during pregnancy, give evidence of perinatal complications, document patterns of twinning, raise suspicions regarding fetal anomalies or explain the source of perinatal diseases, anomalies or growth retardation. The findings ascertained by such examinations may be particularly significant in the neonatal management of the sick infant and genetic counselling of a family delivered of an abnormal fetus. There are many instances where lesions are seen in the placenta without apparent explanation or where the clinical situation would suggest severe placental pathology but none is identified. These frustrating situations may yield in the future to closer clinical pathological correlation and more sophisticated analyses as long as these are undertaken in the context of the two individuals reacting through the placenta.

References

1   Abramowsky, C.R.; Vegas, M.E.; Swinehart, G.; Gyves, M.: Decidual vasculopathy of the placenta in lupus erythematosus. New Engl. J. Med. *303:*668–672 (1980).

2   Baldwin, V.J.; Kalousek, D.K.; Dimmick, J.E.; Applegarth, D.A.; Hardwick, D.F.: Diagnostic pathologic investigation of the malformed conceptus. Perspect. pediatr. Pathol. (in press).
3   Baldwin, V.J.; MacLeod, P.J.; Benirschke, K.: Placental findings in alcohol abuse in pregnancy. A. Rev. Birth Defects, 1981 (in press).
4   Benirschke, K.; Driscoll, S.G.: The pathology of the human placenta, pp. 28–34 (Springer, New York 1967).
5   Benirschke, K.; Gille, J.: Placental pathology and asphyxia; in Gluck, Intrauterine asphyxia and the developing fetal brain, pp. 117–136 (Year Book Medical Publishers, Chicago 1977).
6   Benzie, R.J.: Amniocentesis, amnioscopy and fetoscopy. Clin. Obstet. Gynec. *7:* 439–460 (1980).
7   Blanc, W.A.: Pathology of the placenta, membranes, and umbilical cord in bacterial, fungal and viral infections in man; in Naeye, Kissane, Kaufman, Perinatal diseases. Int. Acad. Path. Monogr., No. 22 (Williams & Wilkins, Baltimore 1981).
8   Brewer, J.I.; Mazur, M.T.: Gestational choriocarcinoma. Am. J. surg. Pathol. *5:* 267–277 (1981).
9   Costoya, A.L.; Leontic, E.A.; Rosenberg, H.G.; Delgado, M.A.: Morphologic study of placental terminal villi in intrahepatic cholestasis of pregnancy. Histochemistry, light and electron microscopy. Placenta *1:* 361–368 (1980).
10   DeFoort, P.; Dhont, M.; Thiery, M.: Hydatidiform mole combined with fetus: extended diagnostic arsenal. Am. J. Obstet. Gynec. *126:* 1049–1051 (1976).
11   Freese, U.E.: A placental evaluation of drug addiction in pregnancy. J. reprod. Med. *20:* 307–315 (1978).
12   Giacoia, G.P.: Hydrops fetalis (fetal edema) – a survey. Clin. Pediat. *19:* 334–339 (1980).
13   Gruenwald, P.: Introduction – The supply line of the fetus; definitions relating to fetal growth; in Gruenwald, The placenta and its maternal supply line, pp. 1–17 (University Park Press, Baltimore 1975).
14   Honoré, L.H.; Dill, F.J.; Poland, B.J.: Placental morphology in spontaneous human abortuses with normal and abnormal karyotypes. Teratology *14:* 151–166 (1976).
15   Hoyme, H.E.; Higginbottom, M.C.; Jones, K.L.: Vascular etiology of disruptive structural defects in monozygotic twins. Pediatrics *67:* 288–297 (1981).
16   Iffy, L.: Ectopic pregnancy; in Iffy, Kaminetzky, Principles and practice of obstetrics and perinatology, pp. 609–633 (Wiley, New York 1981).
17   Jones, C.J.P.; Fox, H.: An ultrastructural and ultrahistochemical study of the human placenta in maternal pre-eclampsia. Placenta *1:* 61–76 (1980).
18   Kaplan, C.; Benirschke, K.: The acardiac anomaly. New case reports and current status. Acta Genet. med. Gemell. *28:* 51–59 (1979).
19   Kitzmiller, J.L.; Cloherty, J.P.; Younger, M.D.; Tabatabaii, A.; Rothchild, S.B.; Sosenko, I.; Epstein, M.F.; Singh, S.; Neff, R.K.: Diabetic pregnancy and perinatal morbidity. Am. J. Obstet. Gynec. *131:* 560–579 (1978).
20   Mannino, F.L.; Jones, K.L.; Benirschke, K.: Congenital skin defects and fetus papyraceus. J. Pediat. *91:* 559–564 (1977).
21   Miller, E.; Hare, J.W.; Cloherty, J.P.; Dunn, P.G.; Gleason, R.E.; Soeldner, J.S.; Kitzmiller, J.L.: Elevated maternal hemoglobin A_{1C} in early pregnancy and major

congenital anomalies in infants of diabetic mothers. New Engl. J. Med. *304:* 1331–1334 (1981).

22 Navarro, C.; Blanc, W.A.: Subacute necrotizing funisitis. J. Pediat. *85:* 689–697 (1974).

23 Ossipoff, V.; Hall, B.D.: Etiologic factors in the amniotic band syndrome: a study of 24 patients. Birth Defects, Orig. Article Ser., No. 13, pp. 117–132 (1977).

24 Pagon, R.A.; Stephens, T.D.; McGillivray, B.C.; Siebert, J.R.; Wright, V.J.; Hsu, L.L.; Poland, B.J.; Emanuel, I.; Hall, J.G.: Body wall defects with reduction limb anomalies: a report of 15 cases. Birth Defects, Orig. Article Ser., No. 15, pp. 171–185 (1979).

25 Powell, H.C.; Benirschke, K.; Favara, B.E.; Pflueger, O.H.: Foamy changes of placental cells in fetal storage disorders. Virchows Arch. Abt. A Path. Anat. Histol. *369:* 191–196 (1976).

26 Queenan, J.T.: Amniotic fluid proteins, amniotic fluid amino acids and their clinical significance; in Fairweather, Eskes, Amniotic fluid – Research and clinical application; 2nd ed., pp. 187–208 (Excerpta Medica, New York 1978).

27 Russell, P.: Inflammatory lesions of the human placenta. III. The histopathology of villitis of unknown etiology. Placenta *1:* 227–244 (1980).

28 Scott, J.N.; Adejokum, F.: Placental changes due to administration of diethylstilbestrol (DES). Virchows Arch. Abt. B Cell Path. *34:* 261–267 (1980).

29 Smith, D.W.: Recognizable patterns of human malformation; 2nd ed., pp. 384–385 (Saunders, Philadelphia 1976).

30 Subrt, I.: Reciprocal translocation with special reference to reproductive failure. Hum. Genet. *55:* 303–307 (1980).

31 Tonkin, I.L.; Setzer, E.S.; Ermocilla, R.: Placental chorangioma: A rare cause of congestive heart failure and hydrops fetalis in the newborn. Am. J. Roentg. *134:* 181–183 (1980).

32 Weinstein, H.J.: Congenital leukemia and the neonatal myeloproliferative disorders associated with Down's syndrome. Clin. Haematol. *7:* 147–154 (1978).

33 Wittmann, B.K.; Baldwin, J.J.; Nichol, B.: Antenatal diagnosis of twin transfusion syndrome by ultrasound. Obstet. Gynec., N.Y. *58:* 123–127 (1981).

V.J. Baldwin, MD, FRCP(C), Division of Pediatric Pathology, Department of Pathology, Vancouver General Hospital, 855 West 12th Avenue, Vancouver, BC V5Z 1M9 (Canada)

Morphometric Evaluation[1]

François Teasdale

Perinatal Service and Research Center, Hôpital Sainte-Justine,
Department of Pediatrics, University of Montreal, Que., Canada

Introduction

The placenta is probably the most versatile and essential organ for fetal survival in utero. For example, it serves as lungs, kidneys, liver and even endocrine organ for the fetus during his intrauterine life. The placenta is also equally important in fetal pathology, because if adequate transfer mechanisms across the placenta between maternal and fetal circulations are impaired, fetal life will be in danger. Therefore, among its diversity of roles, the functional efficiency of the placenta can be defined by its function of transfer of materials between maternal and fetal circulations. Since transplacental exchanges of nutrients critically depend on the fetal and maternal blood flows, and on the total area, thickness and composition of the placental membrane [9], the morphometric evaluation of these structures or compartments should provide important information about placental function.

There are a number of complications of pregnancy, extrinsic or intrinsic to the organ, that will result in placental changes and dysfunction by mechanisms not entirely clear. Therefore, in this chapter, placental dysfunction will be evaluated through the quantitative analysis of the placental components which are intimately related to the transfer function of the placenta.

---

[1] Supported in part by grant MA-6503 from the Medical Research Council of Canada, and by a grant from the Canadian Diabetic Association.

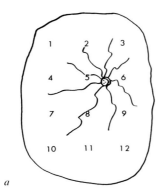 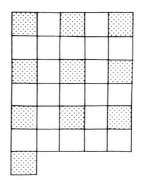

Fig. 1. Schematic representation of the systematic sampling procedures. *a* The numbers on the fetal surface of the placenta indicate the sampling sites. *b* The dotted squares illustrate the pattern of field sampling on the microscopic section.

Materials and Methods

To illustrate the type of information that one can collect from morphometric analyses of the placenta, the results from the study of a placenta of a mother with chronic essential hypertension will be compared with the data collected from 5 normal placentas.

First, it is important to clinically categorize each placenta based on well-defined diagnostic criteria, and with proper control of the same gestational age. The placentas are collected immediately after delivery, the trimmed placental weight is then determined and an estimate of the placental volume is made by the technique of water displacement.

Tissue Sampling and Preparation

Histologic methods of measurement are based on the assumption that the samples on which the observations are made are representative of the whole organ. The human placenta has been shown by previous investigators to be a complex organ, with a pattern of villous variability which is three-dimensional [3–5, 8, 10]. As shown in figure 1a, a systematic sampling procedure is performed, in contrast to a less reliable random technique [13, 14], where 24 samples of tissue are taken at 12 equidistant sites both from the fetal and maternal surfaces of the placenta. These blocks of tissue are therefore good representative samples of the whole organ, both in the horizontal and vertical planes of the placenta. The tissues are then fixed in buffered glutaraldehyde and postfixed in osmium tetroxide, and finally embedded in Araldite [10]. Since the validity of stereological data critically depends on the faithful preservation of cell and tissue dimensions, much consideration is given to the pH and osmolarity of all solutions used for tissue preparation [10]. In contrast to most previous studies which were done on thick (5–8 μm) sections of paraffin-embedded tissues [1, 4, 5, 7, 8], the accuracy of the morphometric measurements performed with the light microscope is enhanced by using sections 1 μm thick cut with an ultramicrotome from these plastic-embedded tissues.

Stereological Analysis

The placenta is a heterogeneous organ, therefore, in order to collect quantitative data that are representative of the transfer function of the placenta, one has to define precisely the structures or compartments that are strictly concerned in the metabolic exchanges between mother and fetus. Accordingly, the parenchymal tissue should be defined as follows [11]: the intervillous space, the trophoblast layer (cytotrophoblast and syncytiotrophoblast) and the fetal capillaries, since they represent the placental membrane separating the fetal and maternal blood compartments, where the transplacental exchanges of materials take place. The nonparenchyma is the sum of decidual and chorionic plates, intercotyledonary septa, fetal vessels, connective tissue of the villi, fibrin deposits and infarcts. The weight and volume of the macrocomponents are determined separately after being dissected with a sharp razor blade [10]. Microscopically, a systematic sampling of fields technique has been designed as shown in figure 1b for the study of each microscopic section [10].

It is now possible with the application of well-known stereological techniques to measure a variety of parameters on cut surfaces of placental tissue. A technique of point-counting, for example, can be applied at the light microscope level by means of a suitable eyepiece graticule to estimate the volume proportions of all microcomponents in the parenchyma and nonparenchyma, based on equation 1 [13, 14]:

$$V_v = \frac{P_i}{P_t} \cdot 100, \tag{1}$$

where P_t represents the number of points on the graticule of the eyepiece and P_i, the total number of hits by the points on the profiles to be analyzed. The absolute weight and volume of each microcomponent as fractions of the placental tissue can then be computed as percentages of the parenchymal and nonparenchymal weight or volume [10]. This point-counting method is also used for the determination of the numerical density of each of the five types of nuclei in the placental tissue. This morphometric parameter is determined according to equation 2 [13, 14]:

$$N_{vn} = \frac{K}{\beta} \cdot \frac{N_{An}^{3/2}}{V_{vn}^{1/2}} \tag{2}$$

in which N_{An} indicates the number of nuclei per unit of area, V_{vn} is the volume fraction of tissue occupied by nuclei, K and β represent the size and shape factors of the particles.

The surface densities of the villi and the fetal vascular bed are measured with a technique of point-intersection counting [13, 14]. The principle is as follows: a graticule line of measured length, projected repeatedly on histological sections, will intersect the boundaries of the component whose surface area we wish to measure. The surface density is computed directly from counts of the intersection points I_i, on the contour of the surface, formed with the test lines of known length L_t:

$$S_v = 2 \cdot \frac{I_i}{L_t}. \tag{3}$$

The numbers of villi, vessels, and capillaries per square centimeter of placental tissue are also recorded, as well as the mean number of syncytiovascular membranes per 100 villi.

Table I. Maternal, fetal and placental characteristics

Characteristic	Control		SGA
	Mean	± SD	
Mother			
Age, years	25.80	3.90	28.00
Weight, kg	69.80	12.64	72.21
Height, cm	160.82	4.75	164.10
Gravidity, n	2.00	0.00	2.00
Blood pressure			
Systolic, mm	106.61	4.22	160.00
Diastolic, mm	68.02	4.47	110.00
Fetus			
Gestational age, weeks	40.00	0.00	40.00
Birth weight, g	3195.01	140.45	1750.00
Birth height, cm	51.13	1.24	45.10
Head circumference, cm	34.10	0.80	30.50
Placenta			
Trimmed weight, g	462.39	68.14	293.94
Trimmed volume, cm^3	438.61	68.59	275.15
Longest diameter, cm	17.92	2.16	15.10
Transverse diameter, cm	18.21	1.30	11.21
Distance of cord implantation from margin, cm	6.51	0.79	4.98
Fibrin, g	11.84	2.71	5.52
Infarcts, g	0.00	0.00	10.13
Parenchyma, g			
Nonparenchyma, g	0.81	0.13	0.43

This is, therefore, a sketch of the morphometric parameters that can be used to investigate the functional structure of the human placenta. The details and the formal mathematical demonstration of the stereological concepts involved in these techniques, which would not be appropriate here, may be filled in by recourse to some of the many publications now available in this field [15, 16].

Results

Table I displays the basic data for the mothers, the infants and their placentas, and figure 2 illustrates the proportions of parenchymal and non-parenchymal tissues in the placentas. The growth characteristics of the newborn infant of the mother with chronic essential hypertension indicate that he is significantly small for gestational age (SGA) compared to the controls

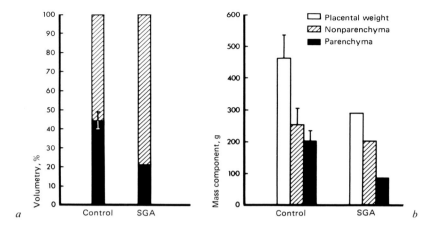

Fig. 2. Parenchymal and nonparenchymal content in relative *(a)* and absolute *(b)* values in the placentas. The placental weight of the SGA infant is significantly smaller than that of the controls, and the ratio of parenchyma to nonparenchyma is markedly reduced (mean ± SD).

and also based on the Montreal intrauterine growth chart [12]. His placental weight is significantly small (293.94 vs. 462.39 g), and the ratio of parenchyma to nonparenchyma is much lower compared to the controls (0.43 vs. 0.81), due mainly to a 57% reduction in parenchymal tissue (88.03 vs. 203.96 g). The distribution of the parenchymal tissue components (intervillous space, trophoblast and fetal capillaries) is illustrated in figure 3. It can be seen that, although their proportions are similar in the placenta of the SGA infant and the controls (fig. 3a), all parenchymal tissue components are markedly decreased in the placenta of the SGA infant (fig. 3b).

The total villous tissue comprises the trophoblast, the connective tissue, the vessels and capillaries of both stem and peripheral villi. Figure 4 demonstrates that there is significantly less villous tissue (83.10 vs. 178.85 g) in the placenta of the mother with chronic hypertension. However, it is noteworthy that this marked difference in villous tissue mass is mainly the result of a 70% reduction in the mass of peripheral villous tissue (45.95 vs. 152.34 g), while the stem villous tissue shows a small increase (37.15 vs. 26.51 g) in the SGA infant's placenta. Accordingly, the peripheral villous and capillary surface areas are significantly decreased, while the stem villous and capillary surface areas are moderately enlarged in the abnormal placenta (fig. 5b, d). In relative values, similar patterns of distribution are illustrated

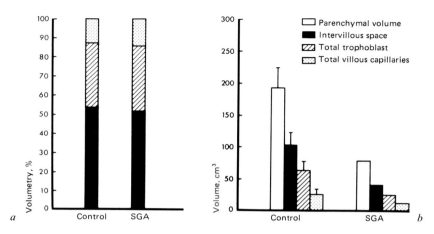

Fig. 3. Percentage distribution *(a)* and absolute volumes *(b)* for the parenchymal components in the placentas. There is a significant reduction in the volume of each of these components in the placenta of the SGA infant.

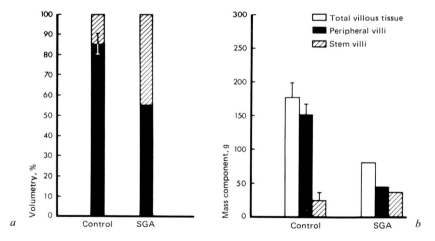

Fig. 4. Distribution in relative *(a)* and absolute *(b)* values between peripheral and stem villous tissues in the placentas. The peripheral villous tissue more than the stem villous tissue is markedly decreased in the placenta of the SGA infant.

Placental Morphometry

Table II. Human placenta: numbers of villi and capillaries per unit area, and the mean number of syncytiovascular membranes per 100 villi

Parameter	Control		SGA
	Mean	± SD	
Number/cm²:			
Stem villi	2,523	678	6,529
Stem villous capillaries	2,597	1,635	7,345
Peripheral villi	32,675	3,840	22,555
Peripheral villous capillaries	50,660	6,411	35,317
Total number of villi	35,198	3,470	29,084
Total number of capillaries	53,257	7,147	42,662
Mean number of syncytiovascular membranes/			
100 villi	146	27	188

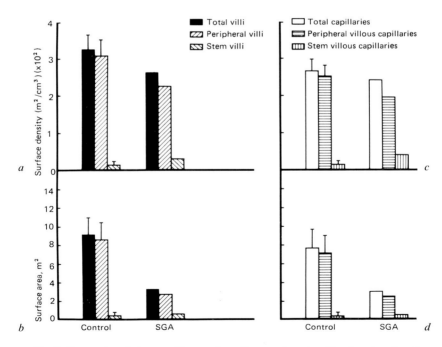

Fig. 5. Comparison between villous and capillary surface densities *(a, c)* and surface areas *(b, d)* in the placentas. The placenta of the SGA infant differs from the controls by having significantly less surface areas of exchange between mother and fetus.

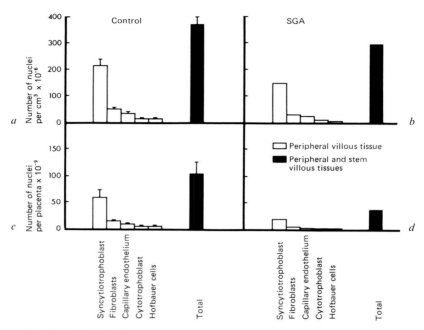

Fig. 6. Comparison between the numerical density *(a, b)* and the actual content per placenta *(c, d)* for the five different types of nuclei in the peripheral villi, and their total number in the peripheral and stem villous tissues. There is a significant decrease in the number of each of the different types of nuclei in the placenta of the SGA infant.

in table II for the number per unit area, and in figure 5 for the surface densities of the peripheral and stem villi and capillaries. There is, however, a tendency for the syncytiovascular membranes to be more numerous (188 vs. 146 per 100 villi; table II) in the villi of the SGA infant's placenta. Figure 6 demonstrates a significant reduction in the relative and absolute numbers of the five different types of nuclei in both peripheral and stem villous tissues of the placenta of the SGA infant compared to the controls.

The purpose of table III is to illustrate the functional significance of these morphological differences between normal placentas and the placental tissue of an intrauterine-growth-retarded infant in relation to fetal weight. The ratios of important parameters of placental function to fetal weight are compared, namely, parenchymal weight, total trophoblast mass, total villous and capillary surface areas, and intervillous space volume. It can be seen that there is a tendency for all these ratios to be lower in the placenta of the

Table III. Human placenta: comparison of different parameters in relation to fetal weight between normal placentas and the placenta of an SGA infant

Parameter	Control		SGA
	Mean	± SD	
Parenchymal weight, g / Fetal weight, g	0.064	0.005	0.050
Total trophoblast weight, g / Fetal weight, g	0.020	0.005	0.016
Total villous surface area, m^2 / Fetal weight, g	2.87×10^{-3}	0.44×10^{-3}	1.88×10^{-3}
Total capillary surface area, m^2 / Fetal weight, g	2.39×10^{-3}	0.51×10^{-3}	1.73×10^{-3}
Intervillous space volume, cm^3 / Fetal weight, g	0.034	0.005	0.024

SGA infant. These results demonstrate a marked reduction in the parenchymal content and in the surface areas of exchange per gram of fetal weight in the SGA infant's placenta compared to normal placental tissues.

Discussion

The quantitative results presented here are based on the assumptions that the samples on which the observations are made are representative of the whole organ, and that cell and tissue dimensions are faithfully preserved. However, even though the described techniques for the preparation of the tissue have been rigorously designed for these types of studies, estimates of intervillous space volume made on a delivered placenta must be cautiously interpreted, for it is likely that it will have collapsed below its natural size in situ. Nevertheless, it seems reasonable to assume that the volumes are measured in placentas which have collapsed by similar proportions on delivery. Therefore, values established on the delivered placenta should, at least, have real comparative worth.

In terms of the transfer function of the placenta, what information can we learn from the measurements of these different parameters? In this typical case of severe intrauterine growth retardation secondary to a maternal disease, we shall concentrate on the placental structures that are intimately involved in the transplacental exchanges of materials between mother and fetus, i.e., the placental membrane, the fetal and maternal blood compartments. The placental membrane in the placenta of the growth-retarded infant is characterized by a 58% decrease in trophoblast mass (28.58 vs. 67.24 g; fig. 3b) compared to the controls. There is also a 67% reduction in the surface area of the peripheral villi (2.85 vs. 8.74 m^2; fig. 5b), the major site of exchange between mother and fetus. Furthermore, in the absence of physiological measurements of blood flows, the results of these quantitative analyses show a 60% reduction in the volume of the intervillous space or maternal blood compartment (41.80 vs. 103.26 cm^3; fig. 3b) and in the surface area (3.03 vs. 7.66 m^2; fig. 5d) of the fetal capillary bed, in the placenta of the SGA infant compared to the controls. These results suggest that the transfer function of the SGA infant's placenta is probably seriously impaired due to this marked reduction in surface areas of exchange between mother and fetus.

Although there is a tendency for the syncytiovascular membranes to be more numerous in the placenta of the growth-retarded infant (188 vs. 146 per 100 villi; table II), it is difficult to imagine that this represents a successful placental adaptation to the maternal disorder, in order to preserve the efficiency of the transfer function of the placenta. Hence, even if it could improve somewhat the transfer of nutrients that are exchanged by simple diffusion, most metabolites are actively transported by the syncytial trophoblast which happens to be significantly reduced in this abnormal placenta (fig. 3b). Therefore morphometry, through the quantitative analysis of the morphological changes which characterize the placenta of a growth-retarded fetus, serves as an important tool to investigate their possible impact on the transfer function of the placenta.

Two other aspects of the functional significance of these morphological differences in the functional structure of the placenta in relation to fetal growth are worth emphasizing: the size of the placenta as a determinant of (1) placental function and (2) fetal growth. These concepts have been thoroughly studied in animals where placental weight at birth and placental function can be compared. In the sheep placenta, for example, it has been shown that, at a given fetal age, despite a process of functional maturation in the course of gestation, placental permeability per kilogram of fetal weight is less than normal when the placenta is small, and thus is proportional to placental weight and DNA content [6]. The functioning mass in terms of parenchymal and cellular contents is markedly decreased in the small placenta of the SGA infant compared to the controls (fig. 2, 6). Therefore, this handicap along with the severely reduced surface areas of exchange between mother and fetus, in terms of capillary and villous surface areas and intervillous space volume (fig. 3, 5), are bound to produce placental dysfunction and fetal growth retardation.

Animal studies have also shown that the placental/fetal weight ratio is often decreased when fetal intrauterine growth retardation occurs associated with a small placenta [2]. Thus, there is evidence of some compensatory growth of the fetus despite the limitation imposed by a small placenta. This concept seems to be also true in humans, as shown in table III, by the lower parenchymal/fetal weight ratio in the placenta of the growth-retarded infant compared to the controls. However, in the absence of physiological data in humans, experimental studies in the ovine fetus have demonstrated that, among fetuses of the same age group, those that are attached to a small placenta, although they are able to achieve a relatively large size in comparison to the weight of their placentas, grow at a rate which is related to the weight, cellular content, and functional capacity of their placentas [6]. Therefore, the lower parenchymal/fetal weight ratio, and the lower ratios of all the surface areas of exchange to fetal weight in the placenta of the growth-retarded infant (table III) imply a narrow margin of safety for the fetus. This may account for the high incidence of fetal distress during the intrapartum period in pregnancies complicated by fetal growth retardation.

The application of this methodology demonstrates that, since physiological studies are difficult to perform in humans without doing any harm to the fetus or the mother, the morphometric analysis of the placenta represents a scientific and noninvasive approach to the field of fetal physiology.

Summary

Placental dysfunction can be evaluated through the quantitative analysis of the morphological changes in the placental structures that are intimately related to the transfer function of the placenta. This is illustrated in the results collected from the placenta of an SGA infant, which show that this placenta differs from a control group by having significantly less functional tissues and surface areas of exchange between mother and fetus. These structural changes can undoubtedly impair placental function and fetal growth in utero. In this context, morphometry is presented as an indirect and noninvasive approach to the study of the physiology and physiopathology of gestation in the human.

References

1 Aherne, W.; Dunnill, M.S.: Quantitative aspects of placental structure. J. Path. Bact. *91:* 123–129 (1966).
2 Alexander, G.: Birth weight of lambs: influences and consequences; in Size at birth. Ciba Fdn Symp., pp. 215–239 (Elsevier, Amsterdam 1974).
3 Alvarez, H.; Benedetti, W.L.; Morel, R.L.; Scavarelli, M.: Trophoblast development gradient and its relationship to placental hemodynamics. Am. J. Obstet. Gynec. *106:* 416–420 (1970).
4 Boyd, P.A.; Brown, R.A.; Stewart, W.J.: Quantitative structural differences within the normal term human placenta: a pilot study. Placenta *1:* 337–344 (1980).
5 Fox, H.: The pattern of villous variability in the normal placenta. J. Obstet. Gynaec. Br. Commonw. *71:* 749–758 (1964).
6 Kulhanek, J.F.; Meschia, G.; Makowski, E.L.; Battaglia, F.C.: Changes in DNA content and urea permeability of the sheep placenta. Am. J. Physiol. *226:* 1257–1263 (1974).
7 Laga, E.M.; Driscoll, S.G.; Munro, H.N.: Comparison of placentas from two socioeconomic groups. I. Morphometry. Pediatrics, Springfield *50:* 24–32 (1972).
8 Mathews, R.; Aikat, M.; Aikat, B.K.: The normal placenta: a histological and histometric study. Indian J, Path. Bact. *16:* 35–40 (1973).
9 Meschia, G.; Battaglia, F.C.; Bruns, P.D.: Theoretical and experimental study of transplacental diffusion. J. appl. Physiol. *22:* 1171–1178 (1967).
10 Teasdale, F.: Functional significance of the zonal morphologic differences in the normal human placenta. Am. J. Obstet. Gynec. *130:* 773–781 (1978).
11 Teasdale, F.: Gestational changes in the functional structure of the human placenta in relation to fetal growth: A morphometric study. Am. J. Obstet. Gynec. *137:* 560–568 (1980).
12 Usher, R.; McLean, F.: Intrauterine growth of live-born Caucasian infants at sea level. J. Pediat. *74:* 901–910 (1969).
13 Weibel, E.R.: Stereological principles for morphometry in electron microscopic cytology. Int. Rev. Cytol. *26:* 235–302 (1969).
14 Weibel, E.R.; Bolender, R.P.: Stereological techniques for electron-microscopic morphometry; in Hayat, Principles and techniques of electron microscopy, vol. 3, pp. 237–296 (Van Nostrand Reinhold Co., New York 1973).
15 Weibel, E.R.: Stereological methods. vol. 1: Practical methods for biological morphometry (Academic Press, New York 1979).
16 Weibel, E.R.: Stereological methods, vol. 2: Theoretical foundations (Academic Press, New York 1980).

Dr. François Teasdale, Perinatal Service, Hôpital Sainte-Justine,
3175, Chemin Sainte-Catherine, Montreal, Que. H3T 1C5 (Canada)

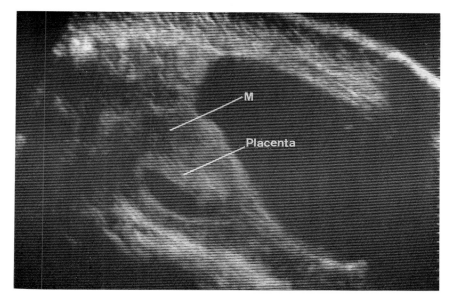

Fig. 2. The same patient as figure 1. The placenta is shown as a thickening of the decidua on the anterior wall of the uterus.

the placentae were refrigerated and subsequently oriented into their intrauterine situation [4].

X-ray examination was performed using a Siemens Pleophos 4 mammograph unit with an Iontomat using 26 to 34 KV at 800 Ma seconds with an exposure of 2 to 4 seconds. The placentae were subsequently photographed before being cut into 1 cm slices corresponding to the levels of the transverse echograms. Further X-rays and photographs were taken.

Findings

In the initial stages of pregnancy the decidua is recognised as a dense white area lining the internal surface of the greyish myometrium and surrounding the echo free amniotic sac (fig. 1). The decidua is normally regular in appearance, but this regularity may be lost if the early pregnancy is jeopardised. *Sample* [unpublished] stated that if the decidual thickness is greater than 5 mm in cases of threatened abortion the prognosis for a successful outcome was high.

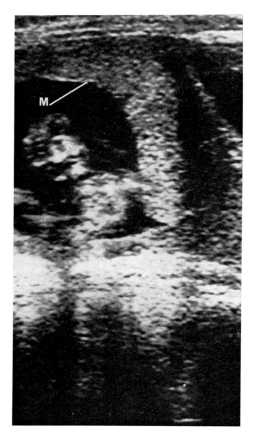

Fig. 3. A longitudinal real time scan at 12 weeks amenorrhoea demonstrating the granular appearance of the placenta and the early appearance of the membranes (M).

As the chorionic villi develop a thickened more heaped, wider area appears which intrudes into the amniotic cavity which is the earliest indication of placental site (fig. 2).

From this time until approximately 36 weeks amenorrhoea the placental volume and surface area will increase as the uterus enlarges although the total area involved relative to the maternal uterine surface does not increase [12]. The placental shape will alter and in some instances its position will appear to change in relation to both intrinsic and extrinsic landmarks within the uterus as the uterine volume enlarges [13].

After 10 weeks amenorrhoea the placental tissue achieves a uniform

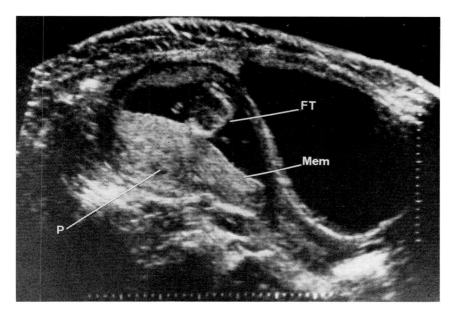

Fig. 4. The appearance of the placenta (P) is now homogeneous. The membranes (Mem) are easily seen. A cross section of the trunk of an 18 week fetus (FT) is shown.

granular appearance and no definite basal layer is apparent [4, 5]. The uniform appearance occurs because the interlobular septa have not yet developed to a state in which significant variations of acoustic impedence occur.

The chorionic plate may be seen as early as 12 weeks [6] (fig. 3) and the amnion and chorion may be separately identified at a later stage, usually 18–22 weeks amenorrhoea [4]. The typical early granular appearance disappears and the placental appearance becomes homogeneous (fig. 4). The chorionic plate is usually smooth. Small echo free spaces are now apparent adjacent to the basal plate which subsequently enlarge and which correspond in location to the sinuses in the chorio-decidual space [4, 10] (fig. 5).

At approximately 28 weeks amenorrhoea small echo free spaces may appear in the placental lobules. These are initially of small diameter (1 mm) and progressively enlarge towards term. The chorionic plate may now show some minor undulations on its surface (fig. 5). *Grannum* et al. [6] have classified placentae of this nature as Grade I and have observed that such an appearance may persist to term. There is still little differentiation of the basal plate.

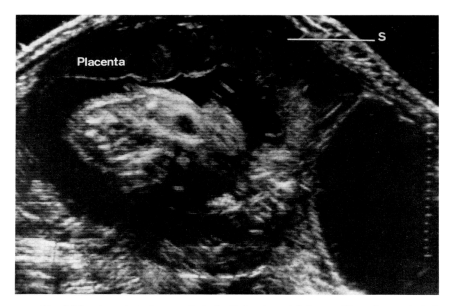

Fig. 5. The placenta is shown anteriorly. Retroplacental sinuses (S) are demonstrated. There is undulation of the chorionic plate.

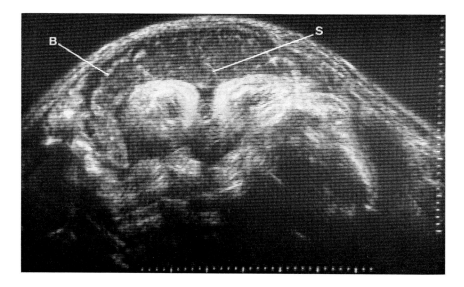

Fig. 6. Basal calcification (B) and interlobular septa (S) seen in a placenta at 28 weeks.

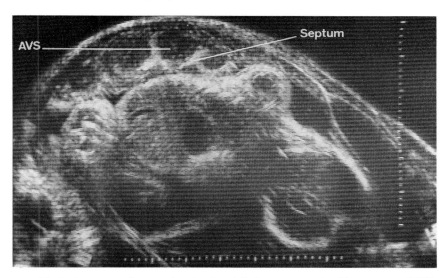

Fig. 7. A mature placenta showing interlobular septae and avillous spaces (AVS).

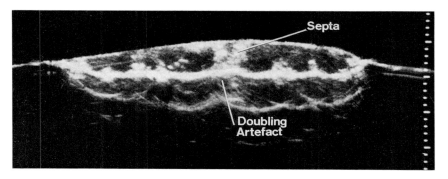

Fig. 8. An echogram of a 1 cm slice from a mature placenta. The septa are well demonstrated as prominent echogenic areas.

As the pregnancy progresses further the echo free areas enlarge in diameter. From 36 weeks onwards they are usually readily apparent. Simultaneously echogenic flecks appear within the placental substance and in relation to the basal plate (fig. 6). The chorionic plate will often show indentations and this appearance has been classified as Grade II [6].

Finally, the echo free areas become obviously separated by densely echogenic tissues running from the chorionic plate deep into the placental

Fig. 9. A radiograph of the placental slice shown in figure 8. The reticular calcification (C) is virtually limited to the basal plate and so cannot be the sole reason for the high level echo's seen in the septa (fig. 8).

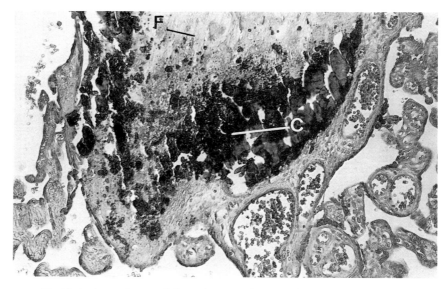

Fig. 10. A photomicrograph from the same placental slice (fig. 8, 9). The interlobular septum is composed basically of fibrin (F) with some calcification (C).

tissue which corresponds to intercotyledonary septa (fig. 7). High echo areas also become apparent along the basal plate. This appearance persists to term and its onset frequently coincides with an observable diminution in the volume of amniotic fluid [4].

Radiological and histological examination has shown that the echogenic areas running from the chorionic plate into the placental substance are due partly to reticular deposits of calcium [4–8] and partly to fibrin formation [11] (fig. 8, 9). The echogenic areas in the basal plate correspond to areas of fine calcification and fibrin deposition (fig. 10). *Grannum* et al. [6] showed that calcification is a major contributing factor to these echogenic areas whilst *Hirabayashi* et al. [14] demonstrated by electron-microscopy that the calcium crystals are deposited on the maternal surface of stem and terminal villi. The intracotyledonary echo free areas have been found to be related to avillous areas located centrally within the lobules [4, 5]. However, there is no uniform agreement on this point [6, 8, 11].

Weinstein and Spirt [15] reported a case of dichorionic diamniotic twins where one twin suffered from growth retardation. Its placenta showed early calcification whilst that of the other twin did not.

Associated Factors

Grannum et al. [6] devised a four stage classification for these changing ultrasonic appearances (0–III). The classification is based on the smoothness of the chorionic plate and the appearance of echo free spaces within the placental substance and the occurrence of echo dense areas both within the placenta and in relation to the basal plate. In 22 patients with Grade III placentae, pulmonary maturation was implied by the finding of a mature lecithin sphingomyelin ratio.

Hirabayashi et al. [14] developed a different classification involving 5 classes (0–IV). They found in their prospective study of 300 patients that 54% of placentae showed signs of calcification. These were classified as III and IV corresponding to *Grannum's* Grade III. In addition they found that placentae showing the most obvious age changes were associated with lower hPL levels than those in whom the changes were less marked. *Iwamoto* et al. [11] devised a similar classification to *Hirabayashi* et al. [14]. However, their Class IV includes calcification in the placental tissue and they claim this is an extension of the Grade III placenta as described by *Grannum* et al. [6]. They postulated that Type IV placentae may equate with diminishing placental reserve and quoted 7 cases of either antepartum or intrapartum fetal distress which occurred in association with Type III or Type IV placentae. Although the numbers involved were only small, the incidence is higher than one would expect in a normal population.

In our original paper [4] a case was described where the occurrence of early maturation changes in the placenta at 32 weeks were associated with

Table I. Incidence of growth retardation

	Hospital deliveries		Early placental maturation changes	
	Total	SGA	Total	SGA
1979	3440	209	36	29
1980	3227	191	27	10

SGA = small for gestational age.

a growth retarded infant. Since then, two retrospective studies have been performed at the Royal Hospital for Women (table I). This shows that in the years 1979 and 1980 the incidence of small for gestational age babies in the general hospital population was 6%. In the 1979 series of patients who showed early maturation changes in their placenta, the incidence of babies who were small for gestational age was 80%, whilst in 1980 37% of babies whose placentas showed early maturation changes were small for gestational age. It is difficult to explain the marked difference in incidence between the two years under study. However, further work is in progress which may shed further light on this problem.

Haney and Traught [8] concluded that transonic placental changes are positively but not unequivocally correlated with intrauterine growth retardation and fetal demise and *Hoogland* [5] concluded that although similar changes were often found in association with infants who were small for gestational age, the results were not statistically significant.

In addition, *Hoogland* [5] has found a slight indication that smoking during pregnancy may influence the early onset of a mature placental appearance.

The Future

Further prospective studies need to be performed to establish whether or not there is any statistically significant correlation between the early appearance of mature placental patterns and infants who are small for gestational age.

Ultrasonic tissue differentiation may be helpful in detecting more subtle changes in placental composition [16] but at this stage there appears to be little correlation using the methods presently available.

Conclusion

Fox [17] has stated that placental calcification is known to be of no pathological or clinical significance, however, because of its high degree of acoustic impedence it is readily recognised in ultrasonic echograms. Its appearance in association with fibrin deposition and their association with enlarging anechoic spaces in the placental substance have been shown to be statistically significant in relation to mature lecithin sphingomyelin ratios. In addition, there is some evidence to suggest that where maturation changes in the placenta occur at an earlier stage than one would expect, that hPL levels may be lower than normal, that there could be a higher incidence of both antepartum and intrapartum fetal distress and that the proportion of small for gestational age babies is higher when compared to a group of babies whose placenta shows an appearance appropriate for its period of gestation.

Acknowledgements

The echograms were obtained in the Department of Diagnostic Ultrasound, Royal Hospital for Women, Paddington (Dr. *W.J. Garrett*, Clinical Director). The figures were prepared by the Department of Medical Illustration, University of New South Wales. Dr. *R.A. Osborn* provided the photomicrograph and the typescript prepared by Mrs. *Lenore Cordin* and Miss *Robyn Bartley*.

References

1   Fox, H.: Pathology of the Placenta, 1st ed., p. viii (W.B. Saunders, London-Philadelphia 1978).
2   Kossoff, G.; Garrett, W.J.; Radovanovich, G.: Grey scale echography in obstetrics and gynaecology. Aust. Radiol. *18:* 62 (1974).
3   Winsberg, F.: Echographic changes with placental ageing. J. Clin. Ultrasound *1:* 52 (1973).
4   Fisher, C.C.; Garrett, W.J.; Kossoff, G.: Placental ageing monitored by grey scale echography. Am. J. Obstet. Gynecol. *124:* 483–488 (1976).
5   Hoogland, H.J.: Ultrasonic aspects of the placenta, 1st ed. (Thume, Nijmegen 1980).
6   Grannum, P.A.T.; Merkowitz, R.H.; Hobbins, J.C.: The ultrasonic changes in the maturing placenta and their relation to fetal pulmonic maturity. Am. J. Obstet. Gynecol. *133:* 915 (1979).
7   Vandenberghe, K.: Placental tissue typing: ultrasonographic and histological correlations. Br. J. Radiol. *53:* 625–630 (1980).

8   Haney, A.F.; Traught, W.S.: Placental appearance in high risk obstetrics: a sonographic correlation. Proceedings 23rd Annual Meeting of the American Institute of Ultrasound in Medicine, *1:* 87 (Mercury Press, 1978).
9   Kawamata, C.; Sugie, T.; Kobayatu, T.F.; Takeuchi, H.; Furuya, H.: Echographic monitoring of placental ageing. Abstracts 2nd Meeting W.F.U.M.B., p. 295 (Scimed, 1979).
10 Spirt, B.A.; Kagan, E.H.; Rozanski, R.M.: Sonographic anatomy of the normal placenta. J. Clin. Ultrasound *7:* 204 (1979).
11 Iwamoto, V.M.; Hashimoto, T.; Tsuzaki, T.; Maeda, K.: Ultrasonographic study of the placenta in vitro. Gynecol. Obstet. Invest. *11:* 265–273 (1980).
12 Vandenberghe, K.: Normal and abnormal placental implantation area: ultrasonographic measurements. Abstracts 2nd Meeting W.F.U.M.B., p. 296 (Scimed, 1979).
13 King, D.L.: Placental migration demonstrated by ultrasonography. Radiology *109:* 167–170 (1973).
14 Hirabayashi, T.; Nakamura, M.; Yokonishi, S.; Horiguchi, T.; Tsukahara, Y.; Kawahara, I.; Fukuta, T.: Study of placental ageing: ultrasonographic and histological correlation. Scientific Proceedings, VIIIth Asian and Oceanic Congress of Obstetrics and Gynaecology, p. 90, Melbourne 1981.
15 Weinstein, H.; Spirt, B.: Sonographic diagnosis of intrauterine growth retardation in a dichorionic diamniotic twin. J. Clin. Ultrasound *7:* 219–221 (1979).
16 Almond, D.C.; Fenton, D.W.; Kennedy, A.; Pryce, W.I.J.: The clinical value of placental tissue characterization using digital techniques. Br. J. Radiol. *54:* 642–648 (1981).
17 Fox, H.: Pathology of the placenta, 1st ed., p. 139 (W.B. Saunders, London-Philadelphia 1978).

C.C. Fisher, FRACOG, Fetal Intensive Care Unit, Royal Hospital for Women, Paddington, N.S.W. (Australia)

Immunopathological Alterations in the Dysfunctioning Placenta

J. Gille

Universitäts-Frauenklinik, Würzburg, BRD

Introduction

How is it possible that the maternal organism tolerates an allogeneic transplant, namely the fetus, for 40 weeks? In pregnancy, all well-known principles of transplantation immunology do not seem to apply. The existence of us all refutes our basic knowledge in this field. But, many questions concerning the enigma of normal pregnancy have been answered so far by careful investigations into immunopathological disorders of pregnancy, like abortion, proliferation of trophoblastic tissue resulting in invasive mole or choriocarcinoma, immunization of the fetus in utero due to blood group incompatibility and EPH gestosis (preeclampsia).

Before discussing the immunopathological alterations of the placenta it may be advantageous to summarize the most recent knowledge on the immunology of uncomplicated pregnancies.

Assuming that pregnant women are *more tolerant to foreign antigens*, maternal lymphocytes were stimulated with streptokinase-streptodornase, tetanus, and *Candida albicans* [Gehrz et al., 1981]. During the third trimester, the cell-mediated response was actually depressed and returned to normal reactivity only after 90 days post partum. The question is open why this tolerance is only demonstrable in the last part of pregnancy. *Youtananoukorn* et al. [1974] could not find any cell-mediated immune reaction in the first trimester. From the 4th month onward, the mother's peripheral leukocytes were reactive to placental antigens. The transformation rate of peripheral lymphocytes to phytohemagglutinin (PHA) dropped in the second and third trimester [*Petrucco* et al., 1976]. Quantitative studies did not show any significant alterations in the circulating T and B lymphocytes in the last trimester [*Cornfield* et al., 1979]. Immunoglobulin-G (IgG)-bearing lymphocytes, however, increased absolutely and relatively. No concept of enhanced immunity should be derived from these observations, however. On the other hand, these studies do not either support the hypothesis of impaired immunity during pregnancy. Obviously, *enhancing antibodies* located in the placenta and suppressor T cells on one side and aggressive sensitized cells on the other side are in a well-balanced state during pregnancy [*Chaouat* et al., 1979].

Autochthonous trophoblastic tissue was attacked by maternal lymphocytes in vitro; the reaction could be reduced by adding maternal or fetal serum [*Taylor*, 1980]. A similar blocking activity could be attributed to a protein isolated from the human syncytiotrophoblast cell membrane when applied to mixed lymphocyte culture (MLC) [*McIntyre and Faulk*, 1979]. No effect was found in PHA-stimulated MLC. In fetal lymphocytes a suppressive effect on maternal cells was present in MLC [*Olding* et al., 1974]. The fetal lymphocytes were considered to belong to the suppressor T cell population [*Olding*, 1979]. The *antigenicity of the trophoblast is reduced* compared to adult tissues. In fetal cell membranes no major histocompatibility complexes were detected [*Parr* et al., 1980]. Histocompatibility and -incompatibility between fetal and maternal cells is one of the most important factors influencing the development of the conceptus. Growth retardation, development of EPH gestosis and abortion are the negative consequences to expect. Before the onset of pregnancy, immunological tolerance of the mother to paternal antigens could improve the chances for the future pregnancy. Considerations have been made concerning spermatozoal histocompatibility antigens [*Marti and Herrmann*, 1977], previous abortions, full-term pregnancies and blood transfusions.

Concluding, pregnancy is made possible by a special state of immunotolerance of the mother triggered by previous contact with paternal antigens. A transplant rejection reaction is prevented by the lack of major histocompatibility complexes of the trophoblastic membrane. Placental antigens, serum factors, diminished cytotoxicity of maternal lymphocytes, blocking antibodies may be additionally helpful in suppressing the maternal immunological response.

Special Clinical Aspects

Abortion
In patients with chronic abortions, absence of a blocking factor in the serum was described [*Rocklin* et al., 1976]. The presence of this factor seems

to be necessary for the maintenance of pregnancy [*Stimson* et al., 1979]. In patients with abortion histories, a normal migration-inhibition reaction was found with autologous antigens. However, autologous plasma did not depress this reaction as it does in non-abortion-prone patients. In a clinical experiment, formation of blocking antibodies was induced by transfusion of leukocyte concentrates from different donors. 4 patients with chronic abortions and no living infants in their history sharing more antigens with their husbands than would be expected statistically had successful pregnancies after this therapy [*Taylor and Faulk*, 1981].

Blocking antibodies cannot be seen morphologically in the placentae of abortions. What can be seen is an increased amount of fibrinoid in the terminal villi [*Gille*, 1981], which could even be found in term placentae with imminent abortion symptoms in early pregnancy [*Ehrhardt* et al., 1972]. Possibly, the increase in fibrinoid can be explained as a hallmark of a more aggressive immunological reaction of the mother against her conceptus.

There will be a more detailed discussion on fibrinoid below (Fibrinoid in the Placenta). Fibrinoid might serve as a protective barrier in the cellular interplay between trophoblast and endometrium [*Wynn*, 1975]. Defects in this fibrinoid layer may release small amounts of antigens which induce maternal tolerance and enable the fetus to survive. Greater amounts of fibrinoid may signalize the impossibility of antigen release from the conceptus, resulting in the lack of blocking antibody formation. Still, there are many open questions about the etiology of recurrent abortions. Hypothetically, even a form of runt disease could hide under the clinical picture of abortion.

Choriocarcinoma

Trophoblastic neoplasia is an overwhelming growth of alloantigens in the maternal organism. In the placental junction zone of invasive hydatidiform mole, the fibrinoid layer was extremely thin, as could be shown by light- and electron-microscopic studies [*Wynn and Harris*, 1967]. Deeper and extensive invasion of trophoblast could be the consequence and one of the reasons of the growth of a trophoblastic tumor. Non-responders to chemotherapy with widespread metastases of choriocarcinoma showed absence of fibrinoid deposits in histological slides, whereas patients with fibrinoid deposits had better responses to therapy and less extensive metastases. Also, the lymphocytic response at the invasion site was more marked in the second group [*Goldstein*, personal commun.]. But the lymphocytic response to carcinoma as a morphological marker for the prognosis [*Bloom and Richardson*, 1957] can no longer be accepted without critical assessment.

Earlier investigations into HL-A compatibility/incompatibility between mother and father in choriocarcinoma suggested histocompatibility between partners to favor trophoblastic neoplasia [*Mogensen and Kissmeyer-Nielsen*, 1968; *Lewis and Terasaki*, 1971]. More recent studies revealed the same normal HL-A distribution in 29 patients with choriocarcinoma and their husbands compared to 21 matched control couples [*Berkowitz* et al., 1981]. All patients with choriocarcinoma showed complete remission with chemotherapy. In a study from Japan, histocompatibility was more remarkable in patients with choriocarcinoma than in those with invasive mole. MLC showed lower reactions in a poor-prognosis group of choriocarcinoma as compared to a good-prognosis group. Stimulation of lymphocytes with PHA was reduced in patients with choriocarcinoma [*Tomoda* et al., 1976]. Histocompatibility between mother and tumor, lowered cellular immunity and lack of fibrinoid at the implantation site seem to be important factors in the etiology of choriocarcinoma and invasive mole. Their effect on the prognosis of the disease has to be considered.

Diabetes mellitus

Placentae in diabetes mellitus have been a favorite subject of morphometric studies. The histological impression of retarded maturation of the villi in the term placenta correlates with a greater surface area combined with a reduced number of syncytial membranes [*Emmrich* et al., 1974]. Immunohistological investigations are rare. *Burstein* et al. [1963] have been the first to apply this technique to diabetic placentae. Using fluorescein-labeled insulin and antihuman globulin they tried to demonstrate specific binding of insulin in proliferative vascular lesions of the placenta and antigen-antibody complexes in the basement membrane as well. For methodological reasons, these studies have to be seen very cautiously. Recent investigations with a more feasible technique detected activation of the complement system in diabetic placentae [*Galbraith* et al., 1981]. Complement components C3 and C4 were increased in the intervillous space and at the trophoblastic basement membrane. The same changes were found in EPH gestosis [*Keane* et al., 1981]. The clotting factors, plasminogen and fibrinogen, were localized in fibrinoid areas. But so far, no interpretation according to the pathogenesis of these alterations has been possible. Additionally, no correlation between the immunohistological findings and the clinical parameters could be established [*Galbraith* et al., 1981].

EPH Gestosis (Preeclampsia)

It is impossible to mention all the theories concerning the etiology of EPH gestosis. In this context, it will be necessary to concentrate on immunological considerations, which does not degrade all the other hypotheses. Higher levels of immune complexes containing IgG with the ability to bind complement were detected in the peripheral blood of preeclamptic patients than in patients with normal pregnancies [*Redman*, 1981]. These complexes obviously are signs of an immunological combat between mother and fetus. Classical essential gestosis occurs in patients who are not preimmunized by contact with paternal antigens, as can be found in spermatozoa, abortions or full-term pregnancies. On the other hand, increased vigilance of the maternal immunological system in rhesus incompatibility decreases the incidence of gestosis.

Histocompatibility between the parents seems to be a predisposing factor for gestosis. *Need* [1975] described a case of uncomplicated first pregnancy in a woman who suffered severe gestosis in her second pregnancy. The second infant's father was different from the first one's and had signs of stronger histoincompatibility with the mother. Similar results were described in a retrospective study by *Feeney* [1980]. Among 34,201 multigravidae, 47 patients developed signs of severe gestosis in a following pregnancy after uncomplicated previous pregnancies. 13 of the 47 patients had different partners in the affected pregnancy, which was significantly higher than in matched controls. A paternal immunogenetic factor is assumed to play an important part in the pathogenesis of gestosis. That maternofetal incompatibility favors gestosis could also be gathered from epidemiological studies from Turkey [*Stevenson* et al., 1976]. Parental consanguinity resulted in a significantly lower incidence of gestoses than nonconsanguinity. Chromosomal aberrations (trisomy 13, 18, 21, XYY syndrome) tend to be a predisposing factor for gestosis [*Müller*, 1978], another finding which confirms the 'incompatibility theory' in the etiology of gestosis. Contradictory results were reported by *Jenkins* et al. [1978]. In 37 patients with severe gestosis, HL-A distribution did not show any difference compared to matched control pairs.

Hyperplacentosis confronts the maternal immune system with an overload of placental antigens provoking the same reactions as genetic dissimilarity: twin pregnancies, hydrops fetalis and hydatidiform moles are combined with a higher incidence of gestoses [*Scott and Beer*, 1976].

The intensity of the maternal immunological response was also tested in cellular studies. With the leukocyte migration inhibition test, 87% positive results were obtained in preeclamptic patients compared to only 22% in women with uneventful pregnancies [*Toder* et al., 1979]. It is assumed that gestosis is a state of cellular hyperreactivity. Blocking antibodies are one of the mechanisms involved in the maintenance of uncomplicated pregnancy [*Stimson* et al., 1979]. Until now, no investigations of blocking antibodies in preeclamptic patients have been performed. Suggesting a deficiency of these antibodies in gestosis, the immunologically more aggressive attitude of the mother against her conceptus could be explained by such studies.

In MLC, fetal T lymphocytes had an inhibiting effect on maternal cells [*Olding*, 1979]. In preeclampsia, maternal lymphocytes are strong enough to resist the fetal inhibitory effect by a significant rise of maternal cells in the two-way MLC [*Gille* et al., 1977].

A higher degree of histoincompatibility and a change of the immunotolerant state of the mother into cellular hyperreactivity in gestosis are correlated with morphological alterations which can be visualized by the fluorescence antibody technique. Changes are localized in the kidney, placenta and uteroplacental vessels.

Renal lesions are characterized by deposits of complement C3, fibrinogen/fibrin and immunoglobulins in glomerular capillaries, the mesangium and arterioles [*Vassalli* et al., 1963; *Petrucco* et al., 1974; *Lichtig* et al., 1975; *Gille*, 1981]. In follow-up studies, months to years after pregnancy, these deposits were no longer present, correlating with the clinical course [*Fiaschi and Naccarato*, 1968; *Fisher* et al., 1977].

Similar deposits have been described in the villous tissue of the placenta and in the walls of the uteroplacental vessels.

Fibrinoid in the Placenta (fig. 1–5). Quantification of fibrinoid deposits in the placentae of preeclamptic patients showed an increased amount of fibrinoid compared to placentae from uncomplicated pregnancies [*Ehrhardt* et al., 1972; *Thliveris and Speroff*, 1977]. These investigations were limited to fibrinoid deposits in the placental villi because fibrinoid in other locations of the placenta is difficult to quantify. But, it has to be kept in mind that fibrinoid in different locations possesses different staining properties [*Burstein* et al., 1973]. It may be concluded that the pathogenesis of fibrinoid varies according to its locations.

Fibrinoid is defined as a substance which is more or less closely related to fibrin [*Boyd and Hamilton*, 1970; *Wynn*, 1975]. Platelet aggregates on the trophoblastic surface can be the first sign of fibrinoid deposition [*Kaufmann*, 1981]. The final total replacement of villous tissue by fibrinoid (fig. 1) re-

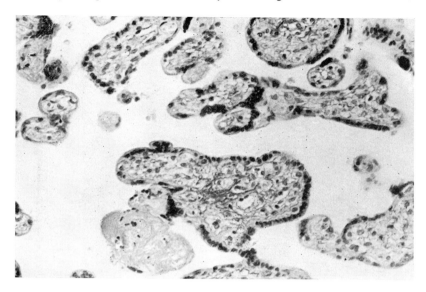

Fig. 1. Total replacement of villous stroma by fibrinoid. Placenta at term. Elastica-van Gieson. × 150.

inforces the placental structure. But, it is very unlikely that fibrinoid is formed for this purpose. This rather represents – teleologically speaking – a not unwelcome side-effect.

By means of immunofluorescence, it was possible to shed more light on the first stages of fibrinoid and thus elucidate its pathophysiological significance. Immunological reactions in fibrinoid deposits were described by *Moe* [1969] and *McCormick* et al. [1971] for the first time with a limited number of antisera. Applying a greater spectrum of fluorescein-conjugated, antihuman antisera, IgG was found in high concentration, IgA and IgM in lower concentrations. Complement C3 and fibrinogen/fibrin were demonstrable to the same degree [*Gille* et al., 1974; *Faulk* et al., 1975]. These deposits were localized on the surface of the villi and in their stroma. Albumin showed different results [*Moe*, 1969; *Gille* et al., 1974]. With special techniques, it was possible to differentiate between fetal and maternal IgG which both contributed to the fluorescence in the fibrinoid areas [*Johnson* et al., 1977]. Distribution and localization of complement components indicated the presence of immune complexes by demonstration of C1q in fetal stem vessels. In fibrinoid material, complement C3c was found [*Faulk*

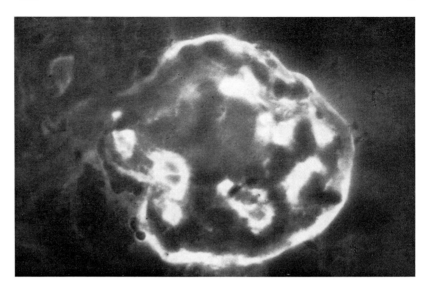

Fig. 2. Fibrinoid deposits on the villous surface and in the stroma. Placenta at term. Incubation with rabbit antihuman fibrinogen (fluorescence antibody technique). ×475.

et al., 1980]. These studies lend support to the possibly necessary effect of complement in the maintenance of normal pregnancy.

In gestosis, the response of the mother, the immunologically mature partner, is stronger than in uncomplicated pregnancies. Since every pregnancy represents an immunological reaction between two allogeneic systems, only quantitative but no qualitative differences can be expected when comparing investigations of fibrinoid deposits. Recently, increased amounts of complement were described at the trophoblastic basement membrane in placentae from patients with gestoses [*Keane* et al., 1981]. Former studies of fibrinoid deposits did not show such differences concerning concentration or composition when placentae from preeclamptic patients and healthy women were compared [*Gille and Halbach*, 1975]. The development of fibrinoid in the placentae of both groups did not reveal any difference either [*Gille* et al., 1974; *Gille and Halbach*, 1975].

The first stage of fibrinoid formation was characterized by deposits of IgG, IgA, IgM, complement C3 and fibrinogen/fibrin on the surface of the villi (fig. 2). The same components were found in the second stage at the trophoblastic basement membrane without any difference in intensity com-

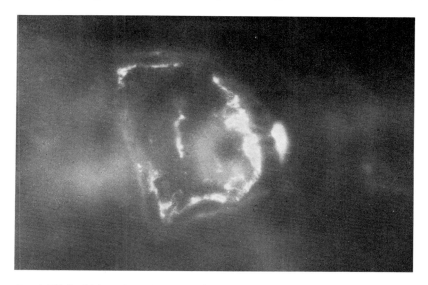

Fig. 3. Fibrinoid deposits at the trophoblastic basement membrane. 32-week placenta. Incubation with goat antihuman complement C3. × 380.

pared to stage one (fig. 3). The combination of deposits on the surface and at the basement membrane is rare; more often, the localization at the basement membrane and in the stroma appeared in the same villus (fig. 4, 5). The deposits in the stroma of the villi, with partial or total replacement (fig. 5) and usually morphologically intact trophoblastic epithelium, contained the already described constituents. Because of their identical composition, the deposits localized at various sites of the villi were suggested to represent different stages in the pathogenesis of the same substance [*Gille*, 1981]. With the PAS stain, positive reactions were present in the thickened trophoblastic membrane and fibrinoid deposits. Additionally, in gestosis, the correlation of these two lesions was significantly higher [*Sen and Langley*, 1974]. These findings are in excellent accordance with the immunohistological alterations described above.

Histochemically, fibrinoid of the villi possesses remarkable similarities with amyloid. Staining with hematoxylin-eosin, Congo red, crystal violet, PAS and van Gieson gave identical results. Birefringence in polarized light and yellow fluorescence after staining with thioflavin T were present in fibrinoid and amyloid [*Beneke* et al., 1970; *Burstein* et al., 1973]. Since amyloid can be considered a result of chronic antigenic stimulation in long-

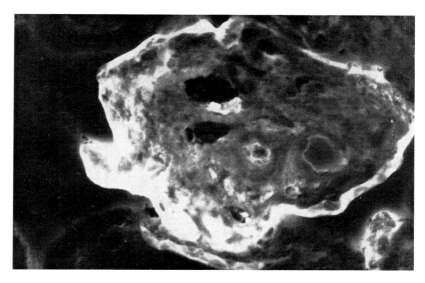

Fig. 4. Fibrinoid deposits on the villous surface and in the stroma, advanced stage. 32-week placenta. Incubation with goat antihuman complement C3. ×475.

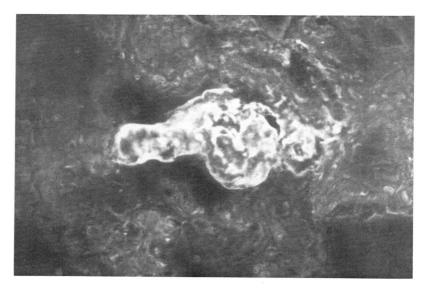

Fig. 5. Fibrinoid deposits with total replacement of the stroma. Placenta at term. Incubation with goat antihuman complement C3. ×237.

lasting diseases or in the aging organism, there are not only similarities with fibrinoid in staining qualities but also in their pathogenetic concept.

Uteroplacental Vessels. The first contact between the trophoblast and its host takes place at the implantation site. The branched tree of the placental villi offers a huge exchange area with the maternal blood in the intervillous space. Fibrinoid deposits are witnesses of the immunological reaction at this site. The implantation, however, is made possible by a much closer contact: the invasion of trophoblast into maternal vessels. These arteries in the placental bed undergo physiological alterations which enable the fetus to get a sufficient maternal blood supply. Distension of the uteroplacental vessels makes them unresponsive to maternal vasomotor agents [*Brosens* et al., 1967]. In gestosis, the distended parts are missing and are replaced by narrow segments [*Robertson* et al., 1967]. Additionally, acute atherosis is present, which is very similar to lesions in rejected kidney transplants [*Brosens*, 1977]. These lesions are very dangerous for the fetus because in gestosis no sufficient adaptation of the placenta to chronic ischemia is available. This is in contrast to essential hypertension where this compensatory possibility improves the chances of the fetus [*Jones and Fox*, 1981].

In normal pregnancies, the walls of the uteroplacental vessels display signs of fibrinogenesis activating the complement system. With the fluorescence antibody technique, 26 of 27 biopsies contained fibrin deposits; additionally, in 12 cases, complement C3 was present [*Weir*, 1981]. In preeclamptic patients, IgG, IgA, IgM, complement C3, C1q and fibrin were found in acute atherosis [*Weir*, 1980]. But, the author is not willing to accept these findings as signs of an antigen-antibody reaction in this disease. *Kitzmiller and Benirschke* [1973] have been able to demonstrate specific deposits of immunoglobulins, complement C3 and fibrin in uteroplacental vessels of preeclamptic women only, while patients with uncomplicated pregnancies had no deposits at all. These results could be confirmed in a greater number of cases, using different antisera [*Gille*, 1979; *Kitzmiller* et al., 1981].

From reduced platelet life span in patients with small-for-gestational age infants it may be concluded that the uteroplacental arteries represent the local site of platelet consumption [*Wallenburg and Kessel*, 1979]. Deterioration of placental perfusion results from the aggregation and formation of thrombi. These alterations in platelet function offer remarkable parallels to the origin of fibrinoid deposition [*Kaufmann*, 1981].

Not only does the light-microscopic finding of acute atherosis resemble alterations found in rejected transplants, but immunofluorescent studies in renal allograft rejection also reveal the same deposits of fibrinogen, complement and immunoglobulins [*Andres* et al., 1970; *Gille* et al., 1970; *McPhaul* et al., 1970] as in the wall of uteroplacental vessels of preeclamptic patients. If identical immunological phenomena may be inferred from identical morphological findings, gestosis should be interpreted as an attempt at an allograft rejection. Cellular hyperreactivity of the mother against her fetus and increased histoincompatibility as a predisposing factor in gestosis are additional facts suitable for this concept.

The mentioned renal lesions could be explained by cross-reactivity with placental antigens, which is readily demonstrable in animal models [e.g. *Gusdon and Witherow*, 1976; *Gille* et al., 1980]. It is very unlikely that cross-reactivity involves other vessels than the kidney. There is only one author who found fibrinoid changes in various parts of the maternal vascular system [*Stoppeli*, 1972]. But others have not been able to confirm these results [*Cretti and Jaworski*, 1975; *Jaworski and Cretti*, 1976]. The livers of preeclamptic patients exhibited deposits resembling those found in the kidneys. In the hepatic sinusoids, fibrin, and to a lesser extent, IgG, IgM and complement C3, were present [*Arias and Mancilla-Jiminez*, 1976].

Placental Size. The influence of immunological factors on placental size has been established in many investigations, with different results. *Beer* et al. [1975] confirmed *James'* [1965] investigations; maternal tolerance seems to be disadvantageous for the development of the placenta. Hybrid placentae of nonsensitized rats, hamsters and mice were heavier than syngeneic placentae. Hybrid placentae of preimmunized animals were heavier than those of nonpreimmunized animals. Hybrid placentae from tolerant animals showed about the same weight as syngeneic placentae. *Hetherington* [1978] could not agree with the deleterious effect of immunotolerance on placental size. In mice, preimmunization against paternal antigens did not reveal any influence on placental size, fetal weight and litter size.

Since there is a statistically significant correlation between placental weight, surface area and fetal size in man [*Gille and Schneider*, 1981; *Emmrich* et al., 1981], it would be of clinical relevance to elucidate immunological factors influencing placental size. In preeclampsia – a state of immunological hyperreactivity – unchanged or larger placentae would be expected from the findings of *Beer* et al. [1975] and *Hetherington* [1978]. Actually, placental weight and surface area were diminished in preeclamptic patients during the 28th and 37th week while 38- to 42-week placentae did not show any difference compared to placentae from uncomplicated pregnancies [*Gille and Halbach*, 1978]. Thus, the immunological influence on placental size remains unconclusive; additionally, it has to be taken into account that the size will also depend on other nonimmunological factors such as genetic constellation, embryonic sex [*Blakley*, 1978] or local alterations at the implantation site from previous operations.

References

Andres, G.A.; Accini, L.; Hsu, K.C.; Penn, I.; Porter, K.A.; Rendall, J.M.; Seegal, B.C.; Starzl, T.E.: Human renal transplants. III. Immunopathological studies. Lab. Invest. *22:* 588–604 (1970).

Arias, F.; Mancilla-Jiminez, R.: Hepatic fibrinogen deposits in pre-eclampsia. Immunofluorescent evidence. New Engl. J. Med. *295:* 578–582 (1976).

Beer, A.E.; Scott, J.R.; Billingham, R.E.: Histoincompatibility and maternal immunological status as determinants of fetoplacental weight and litter size in rodents. J. exp. Med. *142:* 180–196 (1975).

Beneke, G.; Rakow, A.D.; Rakow, L.; Schmitt, W.: Nachweis von Fibrin und Amyloid mit morphologischen Methoden in Gewebsschnitten. Beitr. Path. *141:* 404–427 (1970).

Berkowitz, R.S.; Hornig-Rohan, J.; Martin-Alasco, S.; Klein, S.; Goldstein, D.P.; Bast, R.C.; Wolf, W.C. de: HL-A antigen frequency distribution in patients with gestational choriocarcinoma and their husbands. Placenta, suppl. 3, pp. 263–267 (1981).

Blakley, A.: Maternal and embryonic gene effects on placental weight in mice. J. Reprod. Fertil. *54:* 301–307 (1978).

Bloom, H.J.G.; Richardson, W.W.: Histological grading and prognosis in breast cancer. A study of 1409 cases of which 359 have been followed for 15 years. Br. J. Cancer *11:* 359–377 (1957).

Boyd, J.D.; Hamilton, W.J.: The human placenta (Heffer, Cambridge 1970).

Brosens, I.A.: Morphological changes in the utero-placental bed in pregnancy hypertension. Clin. Obstet. Gynec. *4:* 573–593 (1977).

Brosens, I.; Robertson, W.B.; Dixon, H.G.: The role of the spiral arteries in the pathogenesis of preeclampsia. J. Path. Bact. *93:* 569–580 (1967).

Burstein, R.; Berns, A.W.; Hirata,; Y. Blumenthal, H.T.: A comparative histo- and immunopathological study of the placenta in diabetes mellitus and in erythroblastosis fetalis. Am. J. Obstet. Gynec. *86:* 66–76 (1963).

Burstein, R.; Frankel, S.; Soule, S.D.; Blumenthal, H.T.: Aging of the placenta: autoimmune theory of senescence. Am. J. Obstet. Gynec. *116:* 271–276 (1973).

Chaouat, G.; Voisin, G.A.; Escalier, D.; Robert, P.: Facilitation reaction (enhancing antibodies and suppressor cells) and rejection reaction (sensitized cells) from the mother to the paternal antigens of the conceptus. Clin. exp. Immunol. *35:* 13–24 (1979).

Cornfield, D.B.; Jencks, J.; Binder, R.A.; Rath, C.E.: T and B lymphocytes in pregnant women. Obstet. Gynec., N.Y. *53:* 203–206 (1979).

Cretti, A.; Jaworski, S.: Microscopical studies of myometrial arterioles outside of placental bed in EPH-gestosis. 5th Int. Symp. on EPH Gestosis, Prague 1975. ICRS *3:* suppl. 2, p. 1 (1975).

Ehrhardt, G.; Gerl, D.; Wasmund, B.: Morphometrische Untersuchungen der Frühgeborenenplazenta unter besonderer Berücksichtigung der Mikrofibrinoidablagerungen. Zentbl. Gynäk. *94:* 1110–1115 (1972).

Emmrich, P.; Gödel, E.; Müller, G.: Zur Quantifizierung der diabetischen Reifungsstörung der Plazenta. Pathol. Microbiol. *41:* 253–265 (1974).

Emmrich, P.; Weihrauch, S.; Nachtigal, B.; Winiecki, P.: Grösse der uteroplazentaren Haftfläche und Gewicht der Plazenta im Vergleich zum Gewicht des Neugeborenen. Z. Geburtsh. Perinat. *185:* 161–164 (1981).

Faulk, W.P.; Jarret, R.; Keane, M.; Johnson, P.M.; Boackle, R.J.: Immunological studies of human placentae: complement components in immature and mature chorionic villi. Clin. exp. Immunol. *40:* 299–305 (1980).

Faulk, W.P.; Trenchev, P.; Dorling, J.; Holborow, J.: Antigens on post-implantation placentae; in Edwards, Howe, Johnson, Immunobiology of trophoblast, pp. 113–125 (Cambridge University Press, Cambridge 1975).

Feeney, J.G.: Pre-eclampsia and changed paternity; in Bonnar, MacGillivray, Symonds, Pregnancy hypertension, pp. 41–44 (MTP Press, Lancaster 1980).

Fiaschi, E.; Naccarato, R.: The histopathology of the kidney in toxaemia. Serial renal biopsies during pregnancy, puerperium and several years postpartum. Virchows Arch. Abt. A Path. Anat. *345:* 299–309 (1968).

Fisher, K.; Ahuja, S.; Luger, A.; Spargo, B.H.; Lindheimer, M.D.: Nephrotic proteinuria with pre-eclampsia. Am. J. Obstet. Gynec. *129:* 643–646 (1977).

Galbraith, G.M.P.; Galbraith, R.M.; Paulsen, E.P.: Placental immunopathology in gestational diabetes. Placenta, suppl. 3, pp. 183–192 (1981).

Gehrz, R.C.; Christianson, W.R.; Linner, K.M.; Conroy, M.M.; McCue, S.A.; Balfour, H.H.: A longitudinal analysis of lymphocyte proliferative responses to mitogens and antigens during human pregnancy. Am. J. Obstet. Gynec. *140:* 665–670 (1981).

Gille, J.: Fluoreszenzmikroskopische Untersuchungen an Deziduaarterien bei EPH-Gestose. Arch. Gynaek. *228:* 675 (1979).

Gille, J.: Die Pathogenese der essentiellen EPH-Gestose. Immunologische und morphologische Untersuchungen (Organisation Gestosis Press, Basel 1981).

Gille, J.; Börner, P.; Reinecke, J.; Krause, P.H.; Deicher, H.: Über die Fibrinoidablagerungen in den Endzotten der menschlichen Plazenta. Arch. Gynaek. *217:* 263–271 (1974).

Gille, J.; Halbach, G.: Entstehung von Fibrinoid in der menschlichen Plazenta und seine Bedeutung für die Überlebensfähigkeit des Feten. Fortschr. Med. *93:* 1681–1685 (1975).

Gille, J.; Halbach, G.: Welche Faktoren beeinflussen die Grösse der Plazenta bei EPH-Gestose? In Rippmann, Stamm, EPH-Gestosis, Davos 1977, pp. 184–191 (Organisation Gestosis Press, Basel 1978).

Gille, J.; Schneider, B.: Einfluss der decidualen Haftfläche auf das Gewicht des Neugeborenen am Termin; in Schmidt, Dudenhausen, Saling, Perinatale Medizin. 9. Deutscher Kongress für Perinatale Medizin, Berlin, vol. VIII, pp. 324–325 (Thieme, Stuttgart 1981).

Gille, J.; Stewart, U.; Schneider, B.: The effect of antiserum against guinea pig placenta

on the pregnant guinea pig. 2nd Congr. Int. Soc. Study of Hypertension in Pregnancy, Cairo 1980.
Gille, J.; Williams, J.H.; Hofman, C.P.: The feto-maternal lymphocyte interaction in preeclampsia and in uncomplicated pregnancy. Eur. J. Obstet. Gynec. Reprod. Biol. 7: 227–238 (1977).
Gille, J.; Zobl, H.; Krause, P.H.; Georgii, A.: Immunhistologische Befunde an menschlichen Transplantatnieren. Verh. dt. Ges. Path. 54: 653–654 (1970).
Gusdon, J.P.; Witherow, C.C.: The effect of active immunity against placental proteins on pregnancy in monkeys. Am. J. Obstet. Gynec. 126: 308–312 (1976).
Hetherington, C.M.: Absence of effect of maternal immunization to paternal antigens on placental weight, fetal weight and litter size in the mouse. J. Reprod. Fertil. 53: 81–84 (1978).
James, D.A.: Effects of antigenic dissimilarity between mother and foetus on placental size in mice. Nature, Lond. 205: 613–614 (1965).
Jaworski, S.; Cretti, A.: Comparative microscopic study of the arterioles of omentum, peritoneum, abdominal musculi recti and abdominal aponeurosis and of the arterioles of myometrium in EPH-gestosis. Int. Symp. on Hypertensive Disorders in Pregnancy, Münster 1976.
Jenkins, D.M.; Need, J.A.; Scott, J.S.; Rajah, S.M.; Edwards, J.: HLA and gestosis; in Rippmann, Stamm, EPH Gestosis, Davos 1977, pp. 239–240 (Organisation Gestosis Press, Basel 1978).
Johnson, P.M.; Natvig, J.B.; Ystehede, U.A.; Faulk, W.P.: Immunological studies of human placentae: the distribution and character of immunoglobulins in chorionic villi. Clin. exp. Immunol. 30: 145–153 (1977).
Jones, C.J.P.; Fox, H.: An ultrastructural and ultrahistochemical study of the human placenta in maternal essential hypertension. Placenta 2: 193–204 (1981).
Kaufmann, P.: Fibrinoid; in Becker, Schiebler, Kubli, Die Plazenta des Menschen, pp. 101–111 (Thieme, Stuttgart 1981).
Keane, M.; Hsi, B.L.; Sinha, D.; Faulk, W.P.: Immunological studies of human placentae: complement components in pre-eclamptic chorionic villi. J. Reprod. Immunol. (in press, 1981).
Kitzmiller, J.L.; Benirschke, K.: Immunofluorescent study of placental bed vessels in pre-eclampsia of pregnancy. Am. J. Obstet. Gynec. 115: 248–251 (1973).
Kitzmiller, J.L.; Watt, N.; Driscoll, S.G.: Decidual arteriopathy in hypertension and diabetes in pregnancy: immunofluorescent studies. Am. J. Obstet. Gynec. 141: 773–779 (1981).
Lewis, J.L.; Terasaki, P.I.: HL-A leukocyte antigen studies in women with gestational trophoblastic neoplasms. Am. J. Obstet. Gynec. 111: 547–554 (1971).
Lichtig, C.; Luger, A.M.; Spargo, B.H.; Katz, A.I.; Lindheimer, M.D.: Renal immunofluorescence and ultrastructural findings in preeclampsia. VIth Int. Congr. Nephrol., Firenze 1975. Clin. Res. 23: 368 (1975).
Marti, J.J.; Herrmann, U.: Immunogestosis: a new etiologic concept of 'essential' EPH gestosis with special consideration of the primigravid patient. Preliminary report of a clinical study. Am. J. Obstet. Gynec. 128: 489–493 (1977).
McCormick, J.N.; Faulk, W.P.; Fox, H.; Fudenberg, H.H.: Immunohistological and elution studies of the human placenta. J. exp. Med. 133: 1–18 (1971).

McIntyre, J.A.; Faulk, W.P.: Trophoblast modulation of maternal allogeneic recognition. Proc. natn. Acad. Sci. USA *76:* 4029–4032 (1979).
McPhaul, J.J.; Dixon, F.J.; Brettschneider, L.; Starzl, T.E.: Immunofluorescent examination of biopsies from long-term renal allografts. New Engl. J. Med. *282:* 412–417 (1970).
Moe, N.: Deposits of fibrin and plasma proteins in the normal human placenta. Acta path. microbiol. scand. *76:* 74–88 (1969).
Mogensen, B.; Kissmeyer-Nielsen, F.: Histocompatibility antigens on the HL-A locus in generalised gestational choriocarcinoma. Lancet *i:* 721–724 (1968).
Müller, H.: Chromosomenaberration des Feten als Prädisposition zur Gestose; in Rippmann, Stamm, EPH Gestosis, Davos 1977, pp. 88–90 (Organisation Gestosis Press, Basel 1978).
Need, J.A.: Pre-eclampsia in pregnancies by different fathers: immunological studies. Br. med. J. *ii:* 548–549 (1975).
Olding, L.: Interactions between maternal and fetal/neonatal lymphocytes. Curr. Top. Pathol. *66:* 83–104 (1979).
Olding, L.B.; Benirschke, K.; Oldstone, M.B.A.: Inhibition of mitosis of lymphocytes from human adults by lymphocytes from human newborns. Clin. Immunol. Immunopathol. *3:* 79–89 (1974).
Parr, E.L.; Blanden, R.V.; Tulsi, R.S.: Epithelium of mouse yolk sac placenta lacks H-2 complex alloantigens. J. exp. Med. *152:* 945–955 (1980).
Petrucco, O.M.; Seamark, R.F.; Holmes, K.; Forbes, I.J.; Symons, R.G.: Changes in lymphocyte function during pregnancy. Br. J. Obstet. Gynaec. *83:* 245–250 (1976).
Petrucco, O.M.; Thomson, N.M.; Lawrence, J.R.; Weldon, M.W.: Immunofluorescent studies in renal biopsies in pre-eclampsia. Br. med. J. *ii:* 473–476 (1974).
Redman, C.W.G.: Immunological factors in the pathogenesis of preeclampsia. Contr. Nephrol., vol. 25, pp. 120–127 (Karger, Basel 1981).
Robertson, W.B.; Brosens, I.; Dixon, H.G.: The pathological response of the vessels of the placental bed to hypertensive pregnancy. J. Path. Bact. *93:* 581–592 (1967).
Rocklin, R.E.; Kitzmiller, J.L.; Carpenter, C.B.; Garovoy, M.R.; David, J.R.: Maternal-fetal reaction. Absence of an immunologic blocking factor from the serum of women with chronic abortions. New Engl. J. Med. *295:* 1209–1213 (1976).
Scott, J.R.; Beer, A.E.: Immunological aspects of pre-eclampsia. Am. J. Obstet. Gynec. *125:* 418–427 (1976).
Sen, D.K.; Langley, F.A.: Villous membrane thickening and fibrinoid necrosis in normal and abnormal placentas. Am. J. Obstet. Gynec. *118:* 276–281 (1974).
Stevenson, A.C.; Say, B.; Ustaoglu, S.; Durmus, Z.: Aspects of pre-eclamptic toxaemia of pregnancy, consanguinity, and twinning in Ankara. J. med. Genet. *13:* 1–8 (1976).
Stimson, W.H.; Strachan, A.F.; Shepherd, A.: Studies on the maternal immune response to placental antigens: absence of a blocking factor from the blood of abortion-prone women. Br. J. Obstet. Gynaec. *86:* 41–45 (1979).
Stoppeli, I.: Pathogenese und Therapie der Gestosen. Münch. med. Wschr. *114:* 1190–1196 (1972).
Taylor, P.V.: Maternal and fetal immune responses to human trophoblast antigens. Am. J. Obstet. Gynec. *138:* 293–296 (1980).
Taylor, C.; Faulk, W.P.: Prevention of recurrent abortion with leucocyte transfusions. Lancet *ii:* 68–70 (1981).

Thliveris, J.A.; Speroff, L.: Ultrastructure of the placental villi, chorion laeve, and decidua parietalis in normal and hypertensive pregnant women. Am. J. Obstet. Gynec. *129:* 492–498 (1977).
Toder, V.; Eichenbrenner, I.; Amit, S.; Serr, D.; Nebel, L.: Cellular hyperreactivity to placenta in toxemia of pregnancy. Eur. J. Obstet. Gynec. Reprod. Biol. *9:* 379–384 (1979).
Tomoda, Y.; Fuma, M.; Saiki, N.; Ishizuka, N.; Akaza, T.: Immunologic studies in patients with trophoblastic neoplasia. Am. J. Obstet. Gynec. *126:* 661–667 (1976).
Vassalli, P.; Morris, R.H.; McCluskey, R.T.: The pathogenic role of fibrin deposition in the glomerular lesions of toxemia of pregnancy. J. exp. Med. *118:* 467–478 (1963).
Wallenburg, H.C.S.; Kessel, P.H. van: Platelet lifespan in pregnancies resulting in small-for-gestational age infants. Am. J. Obstet. Gynec. *134:* 739–742 (1979).
Weir, P.E.: An immune reaction in pre-eclampsia? 2nd Congr. of the Int. Soc. for the Study of Hypertension in Pregnancy, Cairo 1980.
Weir, P.E.: Immunofluorescent studies of the uteroplacental arteries in normal pregnancies. Br. J. Obstet. Gynaec. *88:* 301–307 (1981).
Wynn, R.M.: Fine structure of the placenta; in Gruenwald, The placenta and its maternal supply line, pp. 56–79 (MTP Press, Lancaster 1975).
Wynn, R.M.; Harris, J.A.: Ultrastructure of trophoblast and endometrium in invasive hydatidiform mole (chorioadenoma destruens). Am. J. Obstet. Gynec. *99:* 1125–1135 (1967).
Youtananoukorn, V.; Matangkasombut, P.; Osathanondh, D.H.: Onset of human maternal cell-mediated immune reaction to placental antigens during the first pregnancy. Clin. exp. Immunol. *16:* 593–598 (1974).

J. Gille, MD, Universitäts-Frauenklinik, Josef-Schneider-Strasse 4,
D-8700 Würzburg (FRG)

Morphologic Changes in the Hypertensive Placenta[1]

H. Soma, K. Yoshida, T. Mukaida, Y. Tabuchi

Department of Obstetrics and Gynecology, Tokyo Medical College Hospital, Japan

Introduction

Numerous reports have been published on the pathology of the placenta in preeclampsia. There is general agreement that the association of lesions such as infarcts is more frequently found in placentas with toxemia of pregnancy than in controls [50]. Furthermore, that placental infarction is associated with a high risk of perinatal death was supported by the cases in which the placenta demonstrated relatively extensive degrees of infarction [34]. Multiple or serial histologic sections of placental infarcts showed that most uteroplacental arteries exhibited evidence of obstruction, and thrombosis was a frequent finding. Therefore, it was concluded that a placental infarct is a fetal cotyledon which became necrotic due to occlusion of the supplying uteroplacental artery [51]. It has also been observed that acute atherosis of decidual vessels is most commonly associated with placental infarction [55]. Such changes in the decidual vessels are sufficient to produce ischemia of the decidua and adversely affect the placenta.

Many types of placental lesions are significantly more extensive in preeclampsia. These include increased syncytial knots or sprouts, increased numbers of true infarcts and retroplacental hematomas, increased loss of syncytium, proliferation of the cytotrophoblast, thickening of the trophoblastic basement membrane, villous necrosis and acute fibrinoid degeneration of maternal decidual arteries [39], and these pathologic findings are seen more frequently in preeclamptic placentas than in the normal placentas.

---

[1] Supported by a grant from the Ministry of Education of Japan for an ultrastructural study on the materno-fetal relationship.

Studies on the uteroplacental arteries in hypertensive pregnancy and fetal growth retardation have been carried out for many years [13, 21, 27, 42, 44, 45]. *Brosens* et al. [12] demonstrated that a defective development of the placental bed vasculature can be the basis of certain categories of abnormal pregnancy; in cases of albuminuric preeclamptic pregnancy, the spiral arteries do not show the typical replacement of their myometrial segments. Accordingly, the occurrence of acute atherosis in the myometrial segments of the spiral artery in preeclamptic pregnancy can be explained by the defective development of physiological vascular changes and a similar defect can occur in fetal growth retardation in nonhypertensive pregnancies. Thus, hypertensive pregnancy, including preeclampsia, is commonly associated with pathologic changes in the placental bed spiral arteries, resulting in fetal growth retardation [13]. In studies on the uteroplacental arteries in pregnancies complicated by severe fetal growth retardation, which were examined by combined light and electron microscopy, the placentas exhibited extensive infarctions and atheromatous lesions in a high proportion of both the spiral arteries supplying the intervillous space and the basal arteries [44]. In addition, four basic pathological mechanisms consisting of faulty placentation, a reduced mass of functioning villous tissue, abnormal villous development and diffuse damage to the placental villi can be considered as signs of placental malfunction [23]. The pathological features of the placenta in hypertensive pregnancy will be reviewed and discussed in this chapter.

Macroscopic Findings in Hypertensive Placentas

It has been reported that the mean placental weight in cases of preeclampsia tends to be little reduced with increasing degrees of severity of toxemia [22, 29]. The placental/fetal weight ratio in toxemia is often increased [15, 29], although the placental/fetal weight ratio falls evenly.

A series of 194 patients with preeclampsia were classified into a pure type without a previous history of toxemia and a mixed type with superimposed preeclampsia, preexisting essential hypertension or chronic nephritis. The mean placental weight in the pure type was 321.11 ± 122.21 g and the average placental/fetal weight ratio was 0.1526 ± 0.036, while the average placental weight in the mixed type was 325.63 ± 117.60 g, the average placental/fetal weight ratio being 0.1456 ± 0.023. The average fetal weight in the pure type tends to be somewhat higher than that in the mixed

Table I. Placental findings in severe toxemia of pregnancy (53 cases)

Lesions	Pure type		Mixed type		Total, %	Controls (4,764 cases), %
	n	%	n	%		
Infarcts	20	60.6	9	45	54.7	32.8
Intervillous thrombosis	6	18.2	7	35	20.7	10.0
Placenta extrachorialis	7	21.2	7	35	26.4	10.0
Retroplacental hematoma	4	12.1	2	10	11.3	
Marginal hemorrhage	6	18.2	4	20	18.9	2.7
Accessory lobes	7	21.2	3	15	18.9	10.9
Decidua necrosis	6	18.2	3	15	16.9	9.6
Abnormal insertion of the cord	4	12.1	3	15	12.1	1.8
Short cord	4	12.1	4	20	15.1	4.1
Chorionic cyst	1	3	0		1.9	
Chorioangioma	1	3	0		1.9	
Meconium staining	6	18.2	4	20	18.9	12.8

type. In cases with severe preeclampsia, disorders of implantation such as abnormal insertion of the umbilical cord or placenta extrachorialis were often associated [29]. In 150 placentas of intrauterine growth-retarded infants, battledore or velamentous insertion of the cord were found in 23%, and 59% of the placentas were infarcted [16]. On the other hand, the overall incidence of placental infarction is much higher than in uncomplicated pregnancies, rising from about 33% in women with mild preeclampsia to 60% in patients with the severe form of toxemia, and retroplacental hematomas are frequently found (12–15%) in placentas from preeclamptic women [24].

Our data regarding the macroscopic findings in 53 severely toxemic placentas divided into two types are given in table I. Placenta extrachorialis including circumvallation was found in 26.4% (pure type 21.2%, mixed type 35%); abnormal insertion of the cord was found in 12.1% (pure type 12.1%, mixed type 15%). In addition, accessory lobes occurred in 18.9%. These frequencies seem to be higher than those in controls. Among the 53 placentas with severe toxemia, the incidence of placental infarction was 54.7% (60.6% in pure type and 45% in mixed type). The estimate of the infarcts together was less than 5% or more than 15%.

The frequency of intervillous thrombosis was 20.7%, and those of retroplacental hematomas and marginal hemorrhages were 11.3 and 18.9%,

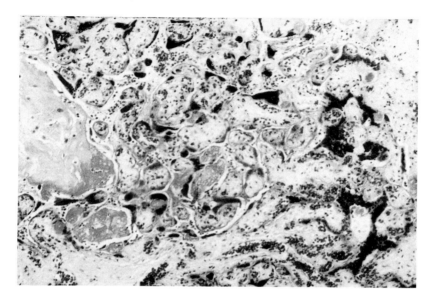

Fig. 1. Microscopic appearance of the acutely infarcted portion of the placenta (case 1). Increased syncytial knots and extensive villous necrosis can be seen. The intervillous space is narrowed and, in some areas, villous capillaries are filled with fetal blood. HE. × 83.

and focal complete degeneration of the trophoblast covered by fibrin thrombi. Placental bed biopsy immediately after removal of the placenta is an excellent way to show lesions of the decidual vessels, which are characteristic for preeclampsia [14, 27]. In general, the physiological changes which normally extend in a retrograde manner down into the myometrial segments of the spiral arteries are restricted to the decidual segments of these arteries, leaving the myometrial segments unaffected. In preeclampsia, acute atherosis is found only in the arteries in the internal zone of the uterine wall and is characterized by fibrinoid necrosis of the vessel wall, presence of lipid and lipophages in the damaged wall and a mononuclear cellular infiltrate around the damaged vessel [12].

Case 1. A 42-year-old gravida 3, para 0 woman referred to our hospital at 23 weeks of gestation because of severe toxemia with hypertension (168/110 mmHg), proteinuria, edema and anemia. Labor was induced with prostaglandin F2α, and a 840-gram stillborn infant was delivered. The placenta weighed 120 g and showed marked marginal infarcts. The edematous umbilical cord was 35 cm long, i.e. too short. Microscopically, many of the lesions clearly showed villous abnormalities due to the severity of the ischemia, which

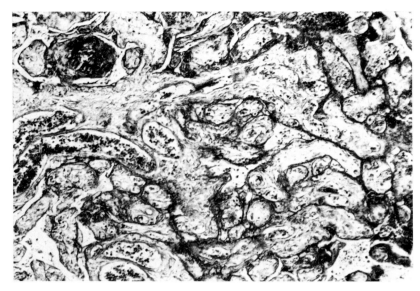

Fig. 2. Microscopically, relatively distended fetal capillaries causing crowding of the villi, narrowing or obliteration of the intervillous space and relative proliferation of the cytotrophoblast can be seen (case 1). Mallory. × 83.

are characteristic features of a damaged syncytium, cytotrophoblastic proliferation, excessive stromal fibrosis, marked formation of syncytial knots and collapse of the villous capillaries. Furthermore, a portion of the intervillous space was obliterated by perivillous fibrin deposition (fig. 1, 2). The wall of the decidual vessel was occupied by marked fibrinoid degeneration including lipid-laden cells, thickening the media of the arterioles and narrowing the lumen (fig. 3).

Case 2. A 30-year-old gravida 2, para 1 woman was admitted with severe toxemia at 33 weeks of gestation, according to our estimate. Her blood pressure was 156/110 mmHg, concomitant with marked proteinuria. She delivered a female infant weighing 1,190 g. The umbilical cord was 32 cm long, i.e. too short. The placenta weighed 245 g and showed marked infarctions as well as intervillous thrombosis associated with a retroplacental hematoma. Microscopically, the most characteristic findings were alterations in the decidual vessels which showed acute atherosis with fibrinoid necrosis of the wall and foamy changes (fig. 4). The villi were characterized by aggregation, fibrinoid necrosis, damaged syncytium with proliferation of syncytial knots and collapsed fetal capillaries.

The histologic features of the hypertensive placenta are summarized in table II. With regard to the pathogenesis of the acute atherosis of preeclampsia later in pregnancy, a feature is the resemblance of the arterio-

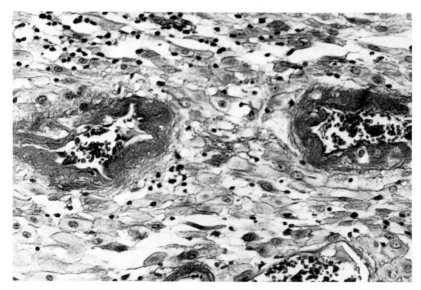

Fig. 3. Moderately thickened and narrowed arterioles with focal fibrinoid degeneration in the basal plate (case 1). HE. ×165.

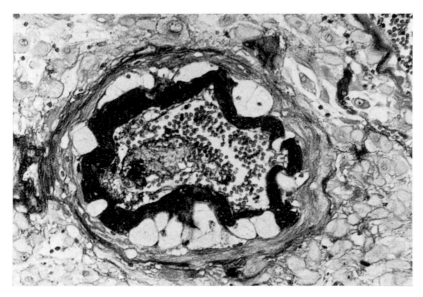

Fig. 4. Acute atherosis with fibrinoid necrosis and foam cells in a decidual artery in case 2. Mallory. ×165.

pathy to a condition suspected to be the result of inappropriate immune response which is seen in rejected renal allografts. In necrotizing arteriopathy, fibrin, IgA, IgM as well as IgG and complement can all be demonstrated by immunofluorescence studies [43]. According to histologic examination and direct immunofluorescence studies on the decidual arteries of the placental bed in hypertensive and diabetic women [33], lesions of fibrinoid necrosis and atherosis were observed in some of the decidual arteries of 53% of women with stable chronic hypertension or normotensive diabetes. Immunoglobulins and complement were detected in lesioned arteries. Probably, in preeclampsia, vascular deposition of immunoglobulin and complement may be related to local intravascular coagulation.

Ultrastructure of the Hypertensive Placenta

Ultrastructural studies on the placental villi in preeclampsia and hypertensive pregnancy have been reported by many investigators [4, 30–32, 41, 48, 49, 54]. The ultrastructural alterations of the placental villi in severe toxemia of pregnancy varies from moderate to severe changes [41]. In severe toxemia of pregnancy, the placental villi exhibit damaged syncytial microvilli showing bulbous swelling or localized loss, dilatation of the syncytial rough endoplasmic reticulum, hydropic swelling of the Golgi apparatus, increase of the cytotrophoblast and thickening of the basement membrane [24, 30, 32] (table II). These studies have more definitely confirmed the histological findings in placental villi in toxemia of pregnancy. According to our investigation, the syncytial microvilli of placental villi in toxemia of pregnancy complicated by hypertension decreased in height with bulbous swelling. The syncytial cytoplasm contained dilated rough endoplasmic reticulum and scattered mitochondria (fig. 5). In some areas, myelin figures were noted, frequently near the nucleus of the syncytiotrophoblast. Platelet and fibrin deposits were in contact with the syncytial surface. The cytoplasm of cytotrophoblastic cells was composed of mitochondria of relatively high density and tubular rough endoplasmic reticulum as well as a large number of free ribosomes. Thickening of the trophoblastic basement membrane was apparent due to its laminated layer appearance. In addition, there was evidence of hypertrophy of the endothelial cells in the fetal capillaries, thereby diminishing the caliber of the lumen.

The ultrastructure of the fetal capillaries of the villi often appears normal despite fibrin deposition on the syncytium, but in most areas where

Table II. Histological and electron-microscopic features of the placenta in hypertensive pregnancy

Site	Microscopic findings	Surface and transmission ultrastructural findings
Chorionic villi and intervillous space	syncytium: focal degeneration, excessive stromal fibrosis, increased syncytial knots, fibrinoid necrosis cytotrophoblast: proliferation of the cytotrophoblast fetal capillary: collapse of the fetal capillaries intervillous space: increased number of infarcts, perivillous fibrin deposits, intervillous thrombosis	villous surface: wrinkled and distorted syncytium: bulbous swelling or localized loss of syncytial microvilli cytotrophoblast: increase of cytotrophoblast fetal capillary: thickening of basement membrane, hypertrophy of endothelial cell, narrowing of the lumen
Decidual artery in the placental bed	atherosis, fibrinoid degeneration, lipid-laden cells (foam cell), narrowing of the arteriolar lumen, mononuclear cell infiltrates around the damaged vessels	fibrin deposits, lipid phage deposits, medial necrosis, myointimal proliferation, luminal thrombosis

this deposition occurs the underlying syncytium lacks microvilli. Extensive fibrin deposition and platelet aggregation may be associated with degeneration of segments of the syncytium where the endothelial cells of the fetal capillaries also show early signs of degeneration [45].

On the other hand, the surface ultrastructure of placentas in severe toxemia of pregnancy exhibited extensive areas of damage in which the surface of the villi was unduly wrinkled and corrugated and showed focal ulceration [26]. According to our observation [36], the surface of terminal villi of toxemic placenta appeared markedly shrunken and distorted, and sometimes agglutinated features could be noted (fig. 6).

Acute atherosis of the myometrial segments of the uteroplacental arteries from hypertensive pregnancy was detected electron-microscopically [19]. At an early stage, the lesion was characterized by endothelial damage, insudation of plasma constituent into the vessel wall, proliferation of myointimal cells and medial necrosis. Ultrastructural studies have also con-

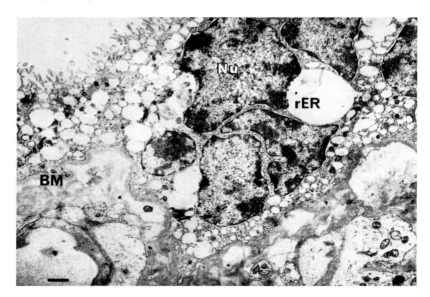

Fig. 5. Chorionic villi in a hypertensive placenta. The syncytial microvilli decrease in height with bulbous swelling. The cytoplasm of the syncytiotrophoblast contains dilated rough endoplasmic reticulum (rER). Thickening of the trophoblastic basement membrane (BM) is consistently observed. Nu = Nucleus. × 2750.

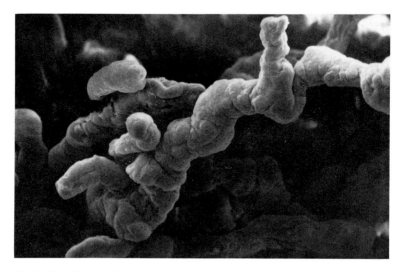

Fig. 6. The villous surface in severe toxemia is unduly shrunken and distorted. × 165.

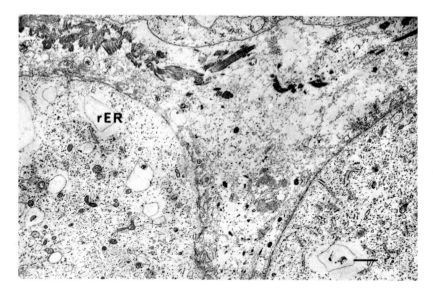

Fig. 7. Beneath the wall of a lesioned decidual artery in hypertensive pregnancy, a fibrin layer and the stroma with edema and regressive changes are seen. Most of the decidual cells exhibit necrotic organelles. rER = Rough endoplasmic reticulum. × 2750.

firmed occlusive vascular lesions including deposition of fibrin or large amounts of lipid and mural thrombi on the wall of the spiral arteries supplying the placenta thus causing severe toxemia or fetal growth retardation [44, 45]. In the luminal wall of the decidual arteries, large amounts of foam cells looking like vacuoles and fibrin deposits were seen; beneath the decidual artery wall, a fibrin layer was also visible, the stroma was edematous, and regressive changes were evident. Most decidual cells showed necrotizing organelles including some vacuoles and mitochondria with disappearing cristae (fig. 7).

On the other hand, it is well known that abnormal coagulation activity involving decrease of platelets, factor VIII consumption, fibrin deposits and fibrinolysis plays a major part in the pathogenesis of both preeclampsia and fetal growth retardation [10, 11]. This would suggest that much more activation of the coagulation system occurs in women with pregnancy-induced hypertension or preeclampsia than in normal pregnancies. The fibrinolytic activity of the intimal cells of decidual spiral arteries was examined by electron microscopy and the fibrin slide technique [46]. This study suggested that in the uteroplacental arteries, the endothelial cells lining the intima of the decidual spiral arteries showed considerably greater fibrinolytic activity than the intimal cytotrophoblast, and the cytotrophoblast in the media reduced fibrinolytic activity in the vessel. Such studies may be useful for confirming the arteriopathic fibrinolytic activity in the uteroplacental junction in hypertensive pregnancies.

Placental Lesions in Experimental Toxemia

In an attempt to produce toxemia experimentally in animals, many studies have been undertaken [1-3, 9, 17, 18, 35, 53], and the placental lesions were described in detail by some investigators [2, 35, 53]. Although some studies relative to the experimental pathogenesis of placental infarction in the rabbit have been reported, it is questionable whether experimental data obtained with the hemochorial but labyrinthine rabbit placenta can be applied to human subjects. Therefore, the rhesus monkey may offer a better experimental model because of the demonstrated close similarity between the fetomaternal circulation in the primate and man [17, 53].

After ligation of uteroplacental arteries in the pregnant rhesus monkey, typical placental infarcts could be demonstrated histologically and histochemically [53]. The earliest lesion was found after a 23-hour interval following ligation. As a result, it was concluded that occlusion of a uteroplacental artery leads to a lesion which is comparable with an infarct in man. On the other hand, when experimental toxemia is produced in the pregnant rabbit by ligating the terminal aorta to a specific degree of stricture, placental lesions such as extensive infarcts with increased syncytial knots resembling those found in human toxemia are observed [2]. However, severe toxemia with similar clinical and pathologic features has been reported to occur spontaneously in patas monkeys and gorillas [5, 8, 28, 40]. A patas monkey spontaneously delivered a stillborn fetus and the placenta was removed manually by hysterotomy. Immediately before delivery, the mother had edematous swelling of the face and abdominal wall as well as oliguria associated with proteinuria. The placenta was typically bilobed, but showed marked infarcts. In the decidual arteries, fibrinoid degeneration and foam cells were seen [28].

A gorilla delivered a stillborn infant. The placenta weighed only 200 g and showed several retroplacental hematomas and marked infarcts. It was concluded that she had suffered from preeclampsia [8]. A multiparous gorilla developed eclampsia before labor and 3 weeks later she delivered a

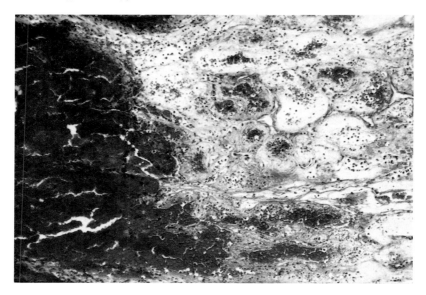

Fig. 8. Marked hemorrhage and necrosis with extensive fibrin thrombosis in the placental labyrinth of the giant panda. HE. ×83.

viable male infant. The placenta weighed 250 g and on gross examination showed fibrotic areas and histologically demonstrated infarcts with syncytial knots and many of the villi with hyaline degeneration [5].

A female giant panda *(Ailuropoda melanoleuca)* suddenly died from uremia with chronic renal insufficiency at Ueno Zoo, Tokyo. At autopsy, this animal was pregnant and had a stillborn fetus. It is possible that this animal was hypertensive, and it is clear that she died from disseminated intravascular coagulation and acute hemorrhagic necrosis; extensive fibrin thrombosis was prominent in the placenta (fig. 8). These pathologic changes not only show important findings of animal experimentation, but are also consistent with placental changes seen in association with maternal hypertension in man.

Summary

Pregnancy complicated by hypertension is commonly associated with placental insufficiency, thereby resulting in fetal growth retardation. Furthermore, reduced uteroplacental blood flow has been recognized in cases of severe preeclampsia with hypertension. Thus, it must be assumed that histological as well as ultrastructural findings in hypertensive placentas are due to the occlusion or narrowing of the uteroplacental vasculature as well as placental ischemia. Microscopically, these placental changes include infarcts, increased syncytial knots, hypovascularity of the villi, cytotrophoblastic proliferation, thickening of the trophoblastic basement membrane, obliterative enlarged endothelial cells in the fetal capillaries and atherosis of the spiral arteries in the placental bed. In addition, ultrastructural features are characterized by a decreased number of syncytial microvilli, proliferation of cytotrophoblastic cells, focal syncytial necrosis, thickening of trophoblastic basement membrane and narrowing of the fetal capillaries, as a number of studies have demonstrated. These placental abnormalities can be seen not only in human toxemia, but also in animals with experimentally induced toxemia or with spontaneous toxemia.

References

1 Abitbol, M.M.; Gallo, G.R.; Pirani, C.L.; Ober, W.B.: Product of experimental toxemia in the pregnant rabbit. Am. J. Obstet. Gynec. *124:* 460–470 (1976).
2 Abitbol, M.M.; Driscoll, S.G.; Ober, W.B.: Placental lesions in experimental toxemia in the rabbit. Am. J. Obstet. Gynec. *125:* 942–948 (1976).
3 Abitbol, M.M.: A simplified technique to produce toxemia in the pregnant dog. Am. J. Obstet. Gynec. *139:* 526–534 (1981).
4 Anderson, W.R.; McKay, D.G.: Electron microscope study of the trophoblast in normal and toxemia placentas. Am. J. Obstet. Gynec. *95:* 1134–1148 (1966).
5 Baird, J.N.: Eclampsia in a lowland gorilla. Am. J. Obstet. Gynec. *141:* 345–346 (1981).
6 Bartholomew, R.A.; Colvin, E.D.; Grimes, W.H., Jr.; Fish, J.S.; Lester, W.M.; Galloway, W.H.: Criteria by which toxemia of pregnancy may be diagnosed from unlabeled formalin-fixed placentas. Am. J. Obstet. Gynec. *82:* 277–290 (1961).
7 Benirschke, K.; Driscoll, S.G.: The pathology of the human placenta (Springer, Berlin 1967).
8 Benirschke, K.; Adams, F.D.: Gorilla disease and causes of death. J. Reprod. Fertil. *28:* suppl., pp. 139–148 (1980).
9 Bergulla, K.; Hennig, F.: Untersuchungen zum Problem des Schwangerschaftshochdrucks als Folge einer utero-placentaren Ischämie. Arch. Gynaek. *225:* 345–351 (1977).
10 Bonnar, J.: The role of coagulation and fibrinolysis in preeclampsia, in Katz, Lindheimer, Zuspan, Hypertension in pregnancy; pp. 85–92 (Wiley, New York 1976).
11 Bonnar, J.: Haemostasis and coagulation disorders in pregnancy; in Bloom, Thomas, Haemostasis and thrombosis, pp. 454–471 (Churchill Livingstone, Edinburgh 1981).
12 Brosens, I.; Dixon, H.G.; Robertson, W.B.: Fetal growth retardation and the arteries of the placental bed. Br. J. Obstet. Gynaec. *84:* 656–663 (1977).
13 Brosens, I.A.: Morphological changes in the utero-placental bed in pregnancy hypertension. Clin. Obstet. Gynec. *4:* 573–593 (1977).
14 Brosens, I.: A study of the spiral arteries of the decidua basalis in normotensive and hypertensive pregnancies. J. Obstet. Gynaec. Br. Commonw. *71:* 222–230 (1964).

15 Budliger, H.: Plazentaveränderungen und ihre Beziehung zur Spättoxikose und perinatalen kindlichen Sterblichkeit. Adv. Obstet. Gynaec., vol. 17, pp. 86–110 (Karger, Basel 1964).
16 Bush, W.: Die Placenta bei der fetalen Mangelentwicklung. Makroskopie und Mikroskopie von 150 Placenten fetaler Mangelentwicklungen. Arch. Gynaek. *212:* 333–357 (1972).
17 Cavanagh, D.; Rao, P.S.; Tung, K.S.K.; Gaston, L.: Eclamptogenic toxemia: the development of an experimental model in the subhuman primate. Am. J. Obstet. Gynec. *120:* 183–196 (1974).
18 Cavanagh, D.; Rao, P.S.; Tsai, C.C.; O'Connor, T.C.: Experimental toxemia in the pregnant primate. Am. J. Obstet. Gynec. *128:* 75–85 (1977).
19 DeWolf, F.; Robertson, W.B.; Brosens, I.: The ultrastructure of acute atherosis in hypertensive pregnancy. Am. J. Obstet. Gynec. *123:* 164–174 (1975).
20 DeWolf, F.; Brosens, I.; Renear, M.: Fetal growth retardation and the maternal arterial supply of the human placenta in the absence of sustained hypertension. Br. J. Obstet. Gynaec. *87:* 678–685 (1980).
21 Dixon, H.G.; Robertson, W.B.: Vascular changes in the placental bed. Pathol. Microbiol. *24:* 622–630 (1961).
22 Diamant, Y.Z.; Kissilevitz, R.: The placenta in intrauterine fetal deprivation. 1. The biochemical profile of placentas from cases of intrauterine growth-retarded gestation of toxemic and non-toxemic origin. Acta obstet. gynec. scand. *60:* 141–147 (1981).
23 Fox, H.: Placental malfunction as a factor in intrauterine growth retardation. in Assche, Robertson, Fetal growth retardation; pp. 117–125 (Churchill Livingstone, Edinburgh 1981).
24 Fox, H.: Pathology of the placenta (Saunders, London 1978).
25 Fox, H.: The histopathology of placental insufficiency. J. clin. Path. *29:* suppl. 10, pp. 1–8 (1976).
26 Fox, H.; Agrafojo-Blanco, A.: Scanning electron microscopy of the human placenta in normal and abnormal pregnancies. Eur. J. Obstet. Gynec. Reprod. Biol. *4/2:* 45–50 (1974).
27 Gerretsen, G.; Huisjes, H.J.; Elema, J.D.: Morphological changes of the spiral arteries in the placental bed in relation to pre-eclampsia and fetal growth retardation. Br. J. Obstet. Gynaec. *88:* 876–881 (1981).
28 Gille, J.H.; Moore, D.G.; Sedgwick, C.J.: Placental infarction: a sign of preeclampsia in a patas monkey *(Erythrocebus patas)*. Lab. Anim. Sci. *27:* 119–121 (1977).
29 Hölzl, M.; Lüthje, D.; Seck-Ebersbach, K.: Placentaveränderungen bei EPH-Gestose. Arch. Gynaek. *217:* 315–334 (1974).
30 Jones, C.J.P.; Fox, H.: An ultrastructural and ultrahistochemical study of the human placenta in maternal pre-eclampsia. Placenta *1:* 61–76 (1980).
31 Jones, C.J.P.; Fox, H.: An ultrastructural and ultrahistochemical study of the human placenta in maternal essential hypertension. Placenta *2:* 193–204 (1981).
32 Kaufmann, P.; Stark, J.: Ultrastruktur der Plazenta bei Diabetes, EPH-Gestose und Rh-Inkompatibilität; in Födisch, Neue Erkenntnisse über die Orthologie und Pathologie der Placenta, pp. 53–62 (Enke, Stuttgart 1977).
33 Kitzmiller, J.L.; Watt, N.; Driscoll, S.G.: Decidual arteriopathy in hypertension and diabetes in pregnancy. Immunofluorescent studies. Am. J. Obstet. Gynec. *141:* 773–779 (1981).

34 Little, W.A.: Placental infarction. Obstet. Gynecol. *15:* 109–130 (1960).
35 McKay, D.G.: The placenta in experimental toxemia of pregnancy. Obstet. Gynec., N.Y. *20:* 1–22 (1962).
36 Mukaida, T.: Surface ultrastructure of the human placenta in abnormal pregnancy (in Japanese). J. Tokyo med. Coll. *39:* 299–310 (1981).
37 Nadji, P.; Sommers, S.C.: Lesions of toxemia in first trimester pregnancies. Am. J. clin. Path. *59:* 344–349 (1973).
38 Naeye, R.: Placental infarction leading to fetal or neonatal death. A prospective study. Obstet. Gynec. *50:* 583–588 (1977).
39 Page, E.W.: On the pathogenesis of pre-eclampsia and eclampsia. J. Obstet. Gynaec. Br. Commonw. *79:* 883–894 (1972).
40 Palmer, A.E.; London, W.T.; Sly, D.L.; Rice, J.M.: Spontaneous preeclamptic toxemia of pregnancy in the Patas monkey *(Erythrocebus patas)*. Lab. Anim. Sci. *29:* 102–106 (1979).
41 Pavelka, M.; Pavelka, R.; Gerstner, G.: Ultrastructure of the syncytiotrophoblast of the human term placenta in EPH-gestosis. Gynecol. obstet. Invest. *10:* 177–185 (1979).
42 Robertson, W.B.: Uteroplacental vasculature. J. clin. Path. *29:* suppl. 10, pp. 9–17 (1976).
43 Robertson, W.B.; Brosens, I.; Dixon, G.: Maternal uterine vascular lesions in the hypertensive complications of pregnancy, in Lindheimer, Katz, Zuspan, Hypertension in pregnancy; pp. 115–127 (Wiley, New York 1976).
44 Sheppard, B.L.; Bonnar, J.: The ultrastructure of the arterial supply of the human placenta in pregnancy complicated by fetal growth retardation. Br. J. Obstet. Gynaec. *83:* 948–959 (1976).
45 Sheppard, B.L.; Bonnar, J.: Ultrastructural abnormalities of placental villi in placentae from pregnancies complicated by intrauterine fetal growth retardation; their relationship to decidual spiral arterial lesions. Placenta *1:* 145–156 (1980).
46 Sheppard, B.L.; Bonnar, J.: Fibrinolysis in decidual arteries in late pregnancy. Thromb. Haemostasis *39:* 751–758 (1978).
47 Soma, H.; Yoshida, K.; Takayama, M.; Tada, M.: Placental tumor (chorioangioma) and its clinical significance (in Japanese). Clin. Gynec. Obstet. *31:* 103–111 (1977).
48 Szarvas, Z.; Treit, P.: Ultrastrukturelle Veränderungen der terminalen Plazentarzotten bei Plazentainsuffizienz. Zentbl. Gynäk. *100:* 1508–1516 (1978).
49 Thliveris, J.A.; Speroff, L.: Ultrastructure of the placental villi, chorion laeve, and decidua parietalis in normal and hypertensive pregnant women. Am. J. Obstet. Gynec. *129:* 492–498 (1977).
50 Wallenburg, H.C.S.: Über den Zusammenhang zwischen Spätgestose und Placentarinfarkt. Arch. Gynaek. *208:* 80–90 (1969).
51 Wallenburg, H.C.S.; Stolte, L.A.M.; Janssens, J.: The pathogenesis of placental infarction. I. A morphologic study in the human placenta. Am. J. Obstet. Gynec. *116:* 835–840 (1973).
52 Wallenburg, H.C.S.: Chorioangioma of the placenta. Thirteen new cases and a review of the literature from 1939 to 1970 with special reference to the clinical complications. Obstetl gynec. Surv. *26:* 411–425 (1971).
53 Wallenburg, H.C.S.; Hutchinson, D.L.; Schuler, H.M.; Stolte, L.A.M.; Janssens, J.:

The pathogenesis of placental infarction. II. An experimental study in the rhesus monkey placenta. Am. J. Obstet. Gynec. *116:* 841–846 (1973).

54 Zacks, S.I.; Blazar, A.S.: Chorionic villi in normal pregnancy, pre-eclamptic toxemia, erythroblastosis, and diabetes mellitus. A light- and electron-microscope study. Obstet. Gynec., N.Y. *22:* 149–167 (1963).

55 Zeek, P.M.; Assali, N.S.: Vascular changes in the decidua associated with eclamptogenic toxemia of pregnancy. Am. J. clin. Path. *20:* 1099–1109 (1950).

Prof. H. Soma, MD, Department of Obstetrics and Gynecology, Tokyo Medical College Hospital, 1-7, Nishishinjuku 6, Shinjuku-ku, Tokyo 160 (Japan)

Vascular Morphology in Diabetic Placentas

Inger Asmussen

Department of Cardiology, University of Copenhagen, Denmark

Vascular disease is a common serious complication of diabetes mellitus. Both large and small vessels will become affected in diabetics, resulting in macro- and microangiopathy. Much effort has been concentrated on identifying the agent(s) responsible for the development of the vascular disease. It is known that diabetic serum induces enhanced smooth muscle cell proliferation [31], a key event in atherogenesis. Potential agents have been identified: hyperlipidemia [31], glucose and sorbitol [54], insulin in vivo [11] and in vitro [52, 53], growth hormone in vivo [69] and in vitro [29, 30], and just recently an agent of low molecular weight present in diabetic serum [28].

During pregnancy, the maternal metabolism will be reflected in the fetal blood. Glucose, free fatty acids and insulin are present in the fetal circulation [8, 45, 50], but at levels below those of the mother; the child has relatively more saturated fatty acids than the mother, the maternal ratio of unsaturated to saturated fat being 1.38 versus 0.68–0.88 in the child [8]. Growth hormone is present in the fetal blood at levels almost four times those of the mother [50]. It seems that the levels of growth hormone are higher in healthy subjects than in diabetics thus indicating that growth hormone alone can not be responsible for the vascular changes found in diabetic placentas.

The fetal glucose levels are higher in diabetics than in healthy subjects [45, 50], corresponding to the higher levels of glucose in the diabetic mother, and thus indicate that the human placenta is readily permeable to glucose, probably reflecting a facilitated diffusion of glucose similar to that of capillaries [43, 45]. Elevated glucose concentrations alter the vascular wall metabolism, reducing the oxygen uptake, increase the tissue water content without affecting the inulin space, inducing increased glycolysis with an increased lactate to pyruvate concentration ratio as well as inducing an increased flux through the polyol pathway [35]. The metabolism of glucose to fructose via sorbitol plays an important role in the development of diabetic complications. It was shown that both enzymes involved in the sorbitol pathway are present in the vascular wall and in the placenta [9, 10, 18, 44]. This change in glucose metabolism, which in mammalian tissue is regarded as an irreversible reaction, combined with reduced hexokinase activity [68] and glycogen accumulation [36], may play an important role in tissue alterations in diabetics. Furthermore, the metabolic changes in diabetics also affect the oxygen transport capacity of the blood [13], with increased levels of glycosylated hemoglobin [6, 20, 34, 38].

Changes in basement membrane material composition occur as a result of high glucose levels similar to the glycosylation of hemoglobin [6, 51], with formation of basement membrane material, with high hydroxylysine, glucose and galactose contents and a decrease in lysine [5, 51]. This biochemical composition may account for the defective filtration function of the capillaries in diabetics and may possibly also alter the placental barrier in diabetics.

During pregnancy, diabetes mellitus will appear with different clinical pictures with various complications in the mother and the child (e.g. birth weight, malformations). *White* [63] subdivided the pregnant diabetics into different groups according to diabetic metabolism and complications. Other classifications [39, 40] have been put forward, but *White's* classification is generally accepted. Similar to these great differences in clinical picture, variability must be expected to be found in the morphology of the placenta. However, only few studies of placental morphology in diabetics have given a proper definition of the material (metabolic state): *Jones and Fox* [26] – gestational diabetes – and *Asmussen* [4] – White's group D, nonsmokers. The reason for excluding smokers from that study [4] was the fact that multilaminal basement membranes and villous fibrosis – similar to the lesions previously ascribed to diabetic metabolism – were found in placentas of smokers [1, 2]. Many of the studies of placental morphology have failed to detect a difference in basement membrane thickening in diabetics compared with controls. Smoking may account for these difficulties. Also the fact that smoking tends to alter villus age so that the terminal villi from heavy smokers appear 'younger' compared with those of controls [2] may give rise to difficulty in the description of material from diabetics (smokers + nonsmokers). Diabetic angiopathy is 4- to 10-fold accelerated in smokers. Thus in the description of the diabetic microangiopathy of the placenta, information on smoking habits is needed.

Diabetic placentas have been studied by light microscopy [7, 14, 15, 17, 24] and electron microscopy [16, 23, 25, 26, 32, 33, 37, 64]. It is generally agreed that increased fibrosis of the villi occurs in combination with signs of dysmaturity. Immature villi with an increased number of cytotrophoblasts are mentioned in several papers [14–17, 24, 25, 64]. However, in all these studies, placentas from patients belonging to various groups according to *White* [63] were described together. This may be the reason why the placentas in the study [4] of White's group D (nonsmokers) were considered to be mature (i.e. very few cytotrophoblasts, pyknotic nuclei of the syncytiotrophoblasts, etc.). That study [4] included a group of controls [1, 2]. The difference in villus maturation can probably be ascribed to a difference in sample collection, as the time of delivery was the same in the various studies (3 weeks prior to estimated term).

Glycogen accumulation is typical of patients with diabetic metabolism. Glycogen deposits are found in various organs: liver, subcutaneous adipose tissue [42], nerves and kidneys [41]. During pregnancy, glycogen storage may occur in the child's liver and heart [21] and in the placenta. Glycogen accumulation is found particularly in the stroma cells and the pericytes surrounding the fetal capillaries but also in the syncytiotrophoblasts [4, 23, 25, 33, 37]. This glycogen accumulation seems to correspond to the finding of increased glycogen content in placentas of diabetic women [12, 55]. The glycogen deposits [4] are mainly located in the stroma cells which often are studded with glycogen granules (in well-controlled diabetics, White's group D!). Smaller deposits were found in the pericytes surrounding the capillaries. Possibly, the cellular glycogen represents deposits, but could likewise be the morphological sign for altered glucose metabolism due either to insulin [36] or to changes in placental enzymes of glycolysis, gluconeogenesis and lipogenesis [12]. In umbilical arteries from White's group D (nonsmokers), similar huge glycogen accumulation was found within the smooth muscle cells [3]. The glycogen accumulation in these arteries was found in proliferating smooth muscle cells and could thus be compared with the lipid accumulation in the proliferating smooth muscle cell in atherogenesis. As large glycogen deposits in the placenta have only been described in association with diabetes mellitus, this seems to be a typical finding of diabetic placentas (fig. 1).

Increased placental residual blood volume is found in diabetics [39]. This might be due to the fact that many diabetics have cesarian section or rather to an increased vascularization of the placenta. At the light-microscopic level, the increased vascularization is visualized: many small capil-

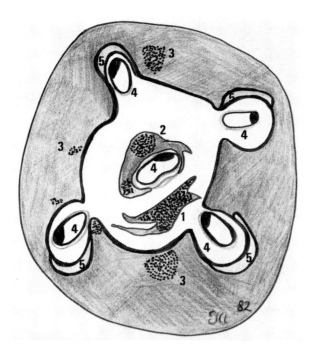

Fig. 1. Schematic drawing showing the localization of the glycogen deposits in diabetic placentas. The largest accumulation is that in the stroma cells (1). Often the cytoplasm of these cells is completely studded with glycogen granules and rosettes. As these cells seem to have a specific affinity for glycogen storage this could possibly represent a true deposit of glycogen in contrast to that of the pericytes (2) and the syncytiotrophoblast (3), where less glycogen is found. The glycogen of these cells (2, 3) may be in response to altered metabolism. 4 = Capillaries; 5 = cytotrophoblast.

laries can be found, mainly located at the periphery of the villus. These small capillaries seem to penetrate deeply into the trophoblast. In many studies of diabetic placentas, this increase in vascularization at the light-microscopic level is described or can be seen in the illustrations [16, 17, 24, 37]. The presence of small capillaries at the villous surface is described by *Lister* [33] and is shown in illustrations from other workers [26, 37]. These vessels seem to be surrounded by sparse basement membrane material and might thus be expected to be newly formed capillaries – a proliferative vascular state [4]. *Emmrich* [16] suggested in his study on basal laminae in diabetic placentas that some villous parts might be younger. This could

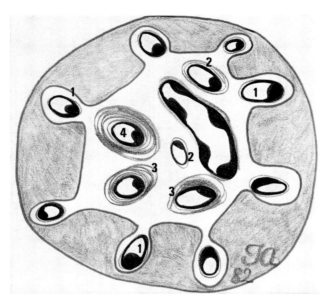

Fig. 2. Schematic drawing showing the microangiopathy of diabetic placentas (the fetal capillaries of the placenta); at the periphery of the villus many small and probably newly formed capillaries are seen penetrating deeply into the trophoblast (1). In the villous stroma, other capillaries are found (2–4). Some of these capillaries (3, 4) exhibit an increased amount of basement membrane material with an increased number of laminae indicating increased endothelial cell turnover similar to that of aging. Thus the diabetic microangiopathy consists of an accelerated aging process in combination with vascular proliferation.

account for his finding of no difference in the width of the basal lamina in diabetics compared with controls. In most studies, diabetic women give birth to their children 3 weeks prior to the estimated term. Therefore, a study on 'healthy' placentas of infants delivered at the same gestational age could establish whether the time of delivery could account for the finding of these peripherally located capillaries in diabetic placentas (fig. 2).

Jones and Fox [25] pointed out that the endothelial cells of the villous capillaries appeared unduly immature. This could also be indicative of a proliferative vascular state. The endothelial cells of the capillaries possessed characteristic cytoplasmic protrusions at the luminal plasma membrane [4]. Similar endothelial cytoplasmic protrusions have been described in umbilical arteries from White's group D (nonsmokers) [3], and can be seen in illustrations of the placental ultrastructure of diabetics [24, 25, 33, 37].

Thickening of the capillary basement membrane is the most typical finding in diabetic microangiopathy. An increase in the width of capillary basal laminae was found both in man and in experimental diabetic animals [27, 46-49, 56, 57, 59-62, 66, 67, 69]. Changes in the width of the capillary basal laminae have been observed in all tissues from diabetics and have also been described in the fetal capillaries of placentas from diabetic women [16, 25, 33, 37]. However, much doubt has been raised as to the validity of the observations of an increase in the width of the basal lamina in diabetics compared with controls. In a morphometric study by *Emmrich* et al. [16], the widths of the capillaries of diabetic placentas were carefully measured and compared with those of a control group. The study showed that the mean width of diabetic capillary basement membranes was less than that of controls. The study tried to explain the 'failure' to show a diabetic angiopathy by the presence of ontogenetically younger villous parts. In fact, this study [16] and a more recent one [4] seem to have pointed out that the diabetic placenta is the site of a proliferative vascular disease. In the study of White's group D placentas, a completely normal thin basal lamina was found surrounding the capillaries (newly formed capillaries) adjacent to capillaries with thick multilaminal basement membrane [4]. *Vracko* [57] and *Vracko and Benditt* [58] explained the appearance of multilaminal basement membrane as a result of repeated endothelial cell death and regeneration. The function of the capillary basal lamina as a scaffold for orderly cell replacement was studied and it was shown that basal laminae increased in number following a number of regenerations [59]. However, that paper [59] also illustrated that an incomplete or only one basal lamina may occur and explained this finding as a result of either removal of old layers of basal lamina by the pericytes or absence of new basal layer formation. Concerning the recent study of *Asmussen* [4], however, one more explanation to the finding of completely normal and thin basal laminae could be the presence of newly formed capillaries. These various events may in part account for the failure to describe a relation between thickening of the basement membrane and the duration/failure of control of the diabetes mellitus. An increase in basement membrane width also occurs with aging [27, 62], and various studies have focused on the relationship between diabetes mellitus and aging [19, 22, 53, 60, 62].

In conclusion, the well-known diabetic microangiopathy develops in the placenta during pregnancy. Proliferation of capillaries in combination with advancing age (multilaminal basement membrane) are typical findings in the capillaries of the terminal villi [65]. The placenta stores glycogen, mainly located in the stroma cells, but also in pericytes and trophoblast. Increased fibrosis and change in maturation are found. Although a great number of studies have already been published, a great effort is needed to study the placenta of carefully selected groups of metabolically well-defined diabetics. Such studies may help to evaluate the clinical situation of the newborn.

Summary

Diabetic metabolism is reflected in the fetal blood and gives rise to diabetic angiopathy in the placenta. Capillary proliferation takes place with small newly formed vessels penetrating deeply into the trophoblast. Also signs of accelerated aging are found with thickening of the capillary basement membrane. Glycogen deposit/storage seems typical of diabetic placenta together with increased fibrosis. Uncertainty exists as to the degree of maturation of the trophoblast.

References

1 Asmussen, I.: Ultrastructure of human placenta at term. Observations on placentas from newborn children of smoking and non-smoking mothers. Acta obstet. gynec. scand. *56:* 119–126 (1977).
2 Asmussen, I.: Ultrastructure of the villi and fetal capillaries in placentas delivered by smoking and non-smoking mothers. II. Br. J. Obstet. Gynaec. *87:* 239–245 (1980).
3 Asmussen, I.: Ultrastructure of human umbilical arteries. Studies on arteries from newborn children delivered by non-smoking, White group D, diabetic mothers. Circulation Res. *47:* 620–627 (1980).
4 Asmussen, I.: Ultrastructure of the villi and fetal capillaries of the placentas delivered by non-smoking diabetic women (White group D). Acta path. microbiol. immunol. scand. sect. A *90:* 95–101 (1982).
5 Beisswenger, P.J.; Spiro, R.G.: Studies on the human glomerular basement membrane. Composition, nature of the carbohydrate units, and chemical changes in diabetes mellitus. Diabetes *22:* 180–193 (1973).
6 Bunn, H.F.; Gabbay, K.H.; Gallop, P.M.: The glycosylation of hemoglobin: relevance to diabetes mellitus. Science *200:* 21–27 (1978).
7 Burnstein, R.; Soule, S.D.; Blumenthal, H.T.: Histogenesis of pathological processes in placentas of metabolic disease in pregnancy. II. The diabetic state. Am. J. Obstet. Gynec. *74:* 96–104 (1957).
8 Chen, C.H.; Adam, P.A.J.; Laskowski, D.E.; McCann, M.L.; Schwartz, R.: The plasma free fatty acid composition and blood glucose of normal and diabetic pregnant women and of their newborns. Pediatrics, Springfield *36:* 843–855 (1965).
9 Clements, R.S., Jr.; Morrison, A.D.; Winegrad, A.I.: Polyol pathway in aorta: regulation by hormones. Science *166:* 1007–1008 (1969).

10 Clements, R.S., Jr.; Reza, M.J.; Winegrad, A.I.: Isolation of the enzymes of the polyol pathway from human placenta. Diabetes *21:* 330 (1972).
11 Cruz, A.B.; Amatuzio, D.S.; Grande, F.; Hay, L.J.: Effect of intra-arterial insulin on tissue cholesterol and fatty acids in alloxan-diabetic dogs. Circulation Res. *9:* 39–43 (1961).
12 Diamant, Y.Z.; Shafrir, E.: Placental enzymes of glycolysis, gluconeogenesis and lipogenesis in the diabetic rat and in starvation. Comparison with maternal and foetal liver. Diabetologia *15:* 481–485 (1978).
13 Ditzel, J.: Oxygen transport impairment in diabetes. Diabetes *25:* 832–838 (1976).
14 Driscoll, S.G.: The pathology of pregnancy complicated by diabetes mellitus. Med. clins N. Am. *49:* 1053–1067 (1965).
15 Emmrich, P.; Amendt, P.; Gödel, E.: Morphologie der Plazenta und neonatale Acidose bei mütterlichem Diabetes mellitus. Pathol. Microbiol. *40:* 100–114 (1974).
16 Emmrich, P.; Fuchs, U.; Heinke, P.; Jutzi, E.; Gödel, E.: The epithelial and capillary basal laminae of the placenta in maternal diabetes mellitus. Lab. Invest. *35:* 87–92 (1976).
17 Fox, H.: Pathology of the placenta in maternal diabetes mellitus. Obstet. Gynec., N.Y. *34:* 792–798 (1969).
18 Gabbay, K.H.: Hyperglycemia, polyol metabolism, and complications of diabetes mellitus. A. Rev. Med. *26:* 521–526 (1975).
19 Goldstein, S.; Littlefield, J.W.; Soeldner, J.S.: Diabetes mellitus and aging: diminished plating efficiency of cultured human fibroblasts. Proc. Acad. Sci. USA *64:* 155–160 (1969).
20 Gonen, B.; Rubenstein, A.H.: Haemoglobin A I and diabetes mellitus. Diabetologia *15:* 1–8 (1978).
21 Halliday, H.L.: Hypertrophic cardiomyopathy in infants of poorly controlled diabetic mothers. Archs Dis. Childh. *56:* 258–263 (1981).
22 Hamlin, C.R.; Kohn, R.R.; Luschin, J.H.: Apparent accelerated aging of human collagen in diabetes mellitus. Diabetes *24:* 902–904 (1975).
23 Hirota, K.; Strauss, L.: Electron microscopic observations on the human placenta in maternal diabetes. Fed. Proc. *23:* 575 (1964).
24 Horký, Z.: Die Reifungsstörungen der Plazenta bei Diabetes mellitus. Zentbl. Gynäk. *87:* 1555–1564 (1965).
25 Jones, C.J.P.; Fox, H.: An ultrastructural and ultrahistochemical study of the placenta of the diabetic woman. J. Path. *119:* 91–99 (1976).
26 Jones, C.J.P.; Fox, H.: Placental changes in gestational diabetes. An ultrastructural study. Obstet. Gynec., N.Y. *48:* 274–280 (1976).
27 Kilo, C.; Vogler, N.; Williamson, J.R.: Muscle capillary basement membrane changes related to aging and to diabetes mellitus. Diabetes *21:* 881–905 (1972).
28 Koschinsky, T.; Bünting, B.; Schwippert, B.; Gries, F.A.: Increased growth stimulation of fibroblasts from diabetics by diabetic serum factors of low molecular weight. Atherosclerosis *37:* 311–317 (1980).
29 Ledet, T.: Growth hormone stimulating the growth of arterial medial cells in vitro. Absence of effect of insulin. Diabetes *25:* 1011–1017 (1976).
30 Ledet, T.: Growth hormone antiserum suppresses the growth effect of diabetic serum. Studies on rabbit aortic medial cell cultures. Diabetes *26:* 798–803 (1977).
31 Ledet, T.; Fischer-Dzoga, K.; Wissler, R.W.: Growth of rabbit aortic smooth-

muscle cells cultured in media containing diabetic and hyperlipemic serum. Diabetes 25: 207–215 (1976).
32 Liebhart, M.: The electron-microscopic pattern of placental villi in diabetes of the mother. Acta med. pol. 12: 133–137 (1971).
33 Lister, U.M.: The ultrastructure of the placenta in abnormal pregnancy. I. Preliminary observations on the fine structure of the human placenta in cases of maternal diabetes. J. Obstet. Gynaec. Br. Commonw. 72: 203–214 (1965).
34 McDonald, J.M.; Davis, J.E.: Glycosylated hemoglobins and diabetes mellitus. Human Pathol. 10: 279–291 (1979).
35 Morrison, A.D.; Clements, R.S.; Winegrad, A.I.: Effects of elevated glucose concentrations on the metabolism of the aortic wall. J. clin. Invest. 51: 3114–3123 (1972).
36 Mulcahy, P.D.; Winegrad, A.I.: Effects of insulin and alloxan diabetes on glucose metabolism in rabbit aortic tissue. Am. J. Physiol. 203: 1038–1042 (1962).
37 Okudaira, Y.; Hirota, K.; Cohen, S.; Strauss, L.: Ultrastructure of the human placenta in maternal diabetes mellitus. Lab. Invest. 15: 910–926 (1966).
38 O'Shaughnessy, R.; Russ, J.; Zuspan, F.P.: Glycosylated hemoglobins and diabetes mellitus in pregnancy. Am. J. Obstet. Gynec. 135: 783–790 (1979).
39 Pedersen, J.: The pregnant diabetic and her newborn. Problems and management; 2nd ed. (Munksgaard, Copenhagen 1977).
40 Pedersen, J.; Mølsted Pedersen, L.: Prognosis of the outcome of pregnancies in diabetics. A new classification. Acta endocr., Copenh. 50: 70–78 (1965).
41 Powell, H.C.; Ward, H.W.; Garrett, R.S.; Orloff, M.J.; Lampert, P.W.: Glycogen accumulation in the nerves and kidney of chronically diabetic rats. J. Neuropath. exp. Neurol. 38: 114–127 (1979).
42 Přibylova, H.; Rážová, M.: Changes of glycogen in subcutaneous adipose tissue and energy metabolism in adaptation period of newborns of normal and diabetic mothers. Czech. Med. 3: 289–294 (1980).
43 Rasio, E.A.: Passage of glucose through the cell membrane of capillary endothelium. Am. J. Physiol. 228: 1103–1107 (1975).
44 Rasio, E.A.; Morrison, A.D.; Winegrad, A.I.: Demonstration of polyol pathway activity in an isolated capillary preparation. Diabetes 21: 330 (1972).
45 Seeds, A.E.; Leung, L.S.; Tabor, M.W.; Russell, P.T.: Changes in amniotic fluid glucose, β-hydroxy-butyrate, glycerol, and lactate concentration in diabetic pregnancy. Am. J. Obstet. Gynec. 135: 887–893 (1979).
46 Siperstein, M.D.: Capillary basement membranes in diabetes; in Fajans, Sussman, Diabetes mellitus: diagnoses and treatment, vol. III, pp. 281–287 (American Diabetes Association, New York 1971).
47 Siperstein, M.D.; Raskin, P.; Burns, H.: Electronmicroscopic quantification of diabetic microangiopathy. Diabetes 22: 514–527 (1973).
48 Siperstein, M.D.; Unger, R.H.; Madison, L.L.: Studies of muscle capillary basement membranes in normal subjects, diabetic and prediabetic patients. J. clin. Invest. 47: 1973–1999 (1968).
49 Siperstein, M.D.; Unger, R.H.; Madison, L.L.: Basement membrane abnormalities in diabetes. Progress in endocrinology, pp. 1136–1141 (Amsterdam, 1969).
50 Spellacy, W.N.; Buhi, W.C.; Bradley, B.; Holsinger, K.K.: Maternal, fetal and amniotic fluid levels of glucose, insulin and growth hormone. Obstet. Gynec., N.Y. 41: 323–331 (1973).

51 Spiro, R.G.: Search for a biochemical basis of diabetic microangiopathy. Claude Bernard Lecture. Diabetologia *12:* 1–14 (1976).
52 Stout, R.W.: Diabetes and atherosclerosis. The role of insulin. Review article. Diabetologia *16:* 141–150 (1979).
53 Stout, R.W.; Bierman, E.L.; Ross, R.: Effect of insulin on the proliferation of cultured primate arterial smooth muscle cells. Circulation Res. *36:* 319–327 (1975).
54 Turner, J.L.; Bierman, E.L.: Effects of glucose and sorbitol on proliferation of cultured human skin fibroblasts and arterial smooth muscle cells. Diabetes *27:* 583–588 (1978).
55 Villee, C.A.: Placenta glycogen metabolism in normal and diabetic subjects; in Camerini-Davalos, Cole, Early diabetes in early life, p. 251 (Academic Press, New York 1975).
56 Vracko, R.: Skeletal muscle capillaries in diabetes. A quantitative analysis. Circulation *41:* 271–283 (1970).
57 Vracko, R.: Basal lamina layering in diabetes mellitus. Evidence for accelerated rate of cell death and regeneration. Diabetes *23:* 94–104 (1974).
58 Vracko, R.; Benditt, E.P.: Capillary basal lamina thickening. Its relationship to endothelial cell death and replacement. J. Cell Biol. *47:* 281–285 (1970).
59 Vracko, R.; Benditt, E.P.: Basal lamina: the scaffold for orderly cell replacement. Observations on regeneration of injured skeletal muscle fibers and capillaries. J. Cell Biol. *55:* 406–419 (1972).
60 Vracko, R.; Benditt, E.P.: Manifestations of diabetes mellitus – their possible relationships to an underlying cell defect. A review. Am. J. Path. *75:* 204–221 (1974).
61 Vracko, R.; Strandness, D.E.: Basal lamina of abdominal skeletal muscle capillaries in diabetics and non-diabetics. Circulation *35:* 690–700 (1967).
62 Vracko, R.; Thorning, D.; Huang, T.W.: Basal lamina of alveolar epithelium and capillaries: quantitative changes with aging and in diabetes mellitus. Am. Rev. resp. Dis. *120:* 973–983 (1979).
63 White, P.: Pregnancy and diabetes. Medical aspects. Med. clins N. Am. *49:* 1015–1024 (1965).
64 Widmaier, G.: Zur Ultrastruktur menschlicher Placentazotten beim Diabetes mellitus. Arch. Gynäk. *208:* 396–406 (1970).
65 Wigglesworth, J.S.: Vascular anatomy of the human placenta and its significance for placental pathology. J. Obstet. Gynaec. Br. Commonw. *76:* 979–989 (1969).
66 Williamson, J.R.; Kilo, C.: Basement-membrane thickening and diabetic microangiopathy. Diabetes *25:* 925–927 (1976).
67 Williamson, J.R.; Kilo, C.: Current status of capillary basement membrane disease in diabetes mellitus. Diabetes *26:* 65–73 (1977).
68 Yalcin, S.; Winegrad, A.I.: Defect in glucose metabolism in aortic tissue from alloxan diabetic rabbits. Am. J. Physiol. *205:* 1253–1259 (1963).
69 Østerby, R.; Seyer-Hansen, K.; Gundersen, H.J.G.; Lundbæk, K.: Growth hormone enhances basement membrane thickening in experimental diabetes. Diabetologia *15:* 487–489 (1978).

I. Asmussen, MD, Department of Cardiology – 2014, Rigshospitalet, University of Copenhagen, DK-2100 Copenhagen (Denmark)

Ultrastructure of Uteroplacental Arteries

F. De Wolf[a], *I. Brosens*[a], *W.B. Robertson*[b]

[a]Department of Obstetrics and Gynaecology, Katholieke Universiteit Leuven, Belgium; [b]Department of Histopathology, St. George's Hospital Medical School, University of London, England

Introduction

Clinical and pathological studies [2, 13] indicate that an insufficient supply of oxygen and essential nutrients to the fetus is most commonly due to impaired maternal blood flow through the intervillous space of the placenta. This implies a defective uteroplacental circulation as the blood supply to the conceptus is ultimately from the spiral arteries, transformed to the uteroplacental arteries in the placental bed. Over the last 20 years several investigators have studied the morphology of the spiral arteries in normal and abnormal pregnancy by light and electron optical techniques. Unfortunately, confusing and contradictory results have been reported and definitions and nomenclature remain to be standardized. This situation has arisen for a variety of reasons: first, the different types of material used, the basal plate of the fetal placenta, placental bed biopsies or hysterectomy specimens; second, failure to appreciate that pregnancy-induced changes in the uterine vessels evolve throughout pregnancy; third, insufficient care taken to differentiate pathological from normal physiological changes in the uteroplacental arteries and, fourth, the attitudes, objectives and, indeed, prejudices of the investigators.

The interpretation of morphological changes in the uteroplacental arteries is particularly difficult in the basal plate, the placental septa and in the superficial layer of the basal decidua because, as pregnancy approaches term, degenerative and senescent changes in fetal and maternal tissue and extensive fibrin deposition produce features that are easily misinterpreted as pathological. Furthermore, basal arteries and branches of spiral arteries that do not communicate with the intervillous space are prone to undergo occlusive changes with consequent decidual necrosis [23]. Tissue sampling problems are another cause of errors in interpretation as the normal pregnancy-induced vascular changes are most extensive in the central area of the placental bed and diminish in extent towards the periphery.

Finally, in studies of abnormal pregnancy, such as the hypertensive disorders and intrauterine fetal growth retardation, it is essential that standardized definitions and classification be adhered to and account be taken of the difficulties of classifying pregnancy disorders in cases where pregnancy is terminated several weeks before term. For example, material derived from a pregnancy terminated at 36 weeks may show features usually associated with preeclampsia but the onset of the clinical features of preeclampsia may have been prevented by the termination of the pregnancy for, say, intrauterine fetal growth retardation.

The Placental Bed Spiral Arteries in Normal Pregnancy

Before discussing acquired lesions of the uteroplacental arteries in abnormal pregnancy, it is necessary to have a clear picture of the structural changes occurring in these vessels during normal pregnancy.

Light Microscopy
Our knowledge of the morphological changes is based mainly on the studies of *Hamilton and Boyd* [14, 15], *Boyd and Hamilton* [3], *Brosens* et al. [5], *Harris and Ramsey* [16], and *Pijnenborg* et al. [23]. All these investigators studied the placental bed spiral arteries in step serial sections of pregnant human uteri with the placenta in situ. The uteri were obtained by hysterectomy at different stages of pregnancy. Their findings can be summarized as follows. During normal pregnancy non-villous migratory trophoblast invades the placental bed, including the lumen of the placental bed spiral arteries. In the latter, there is a striking distension of the vessel lumen and degenerative changes occur in the vessel wall. The musculoelastic elements are replaced by fibrinoid material in which large cells are embedded. As pregnancy proceeds, these changes extend from the decidual portions proximally into the myometrial portions of the spiral arteries and may even involve the terminal segments of the radial arteries. It was the merit of these workers to have recognised and interpreted these changes occurring during normal pregnancy as physiological changes.

Electron Microscopy

De Wolf et al. [9] were the first to apply the technique of electron microscopy in an attempt to identify the nature and the origin of the large intramural cells in the vessel wall and of the fibrinoid material associated with them. Placental bed biopsies, to include basal decidua and underlying myometrium, were taken during Caesarean section at or near term of normal pregnancy.

Nature of the Large Intramural Cells

The decidual segment of the spiral artery is more severely affected by physiological changes than the myometrial segment. The large intramural cells and the fibrinoid material in the myometrial segments differ in no significant way from those seen in the decidual segments of the spiral arteries exhibiting the same variations in structure. The morphology of the intramural cells is not uniform. The cells are often irregular with cytoplasmic processes, surrounded by basement membrane-like substance and frequently connected by desmosomes. The nuclei also tend to be irregular in outline. The endoplasmic reticulum is prominent and of the dilated cisternal type, containing amorphous secretion of low electron density.

There are numerous cytoplasmic filaments in aggregates between abundant endoplasmic reticulum. The ratio of filaments to endoplasmic reticulum varies from cell to cell and within cells. Some cells appear to be transformed to a spindle shape with finely branched long cytoplasmic processes containing few organelles. These variations can be explained on the basis of aging of the cell. The large intramural cells thus have all the features in keeping with mature, and in many cases senescent, trophoblast.

Nature and Origin of the Fibrinoid Material

The 'fibrinoid material' in which the trophoblast is embedded appears as a complex of identifiable fibrin of maternal origin and other fibrillar and granular protein, possibly derived from the cytoplasmic filamentous material of degenerating trophoblast. The presence of degenerate cell products in the 'fibrinoid' would indicate an origin for at least some of the material from this source. The cisternae of the rough endoplasmic reticulum of the intramural trophoblast contain material of low electron density and fusion between these cell organelles and the peripheral plasma membrane is a frequent finding. These observations support the hypothesis that the fibrinoid material is also partly the result of apocrine secretion by the trophoblast. In addition, there are the residua of elastica and ground substance, normal components of an arterial wall.

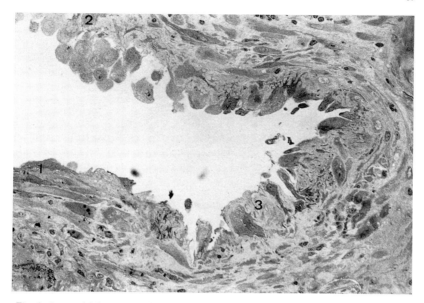

Fig. 1. 1-μm thick epon-embedded section showing a placental bed spiral artery with the three different stages of modifications found in the second trimester of normal human pregnancy. ×110.

The findings of *Sheppard and Bonnar* [29] on the ultrastructure of the physiological changes are similar to our findings.

Trophoblastic Invasion of Spiral Arteries

Different hypotheses based on light optical studies [5, 15, 16] have been formulated as to how the trophoblast becomes intramural. To evaluate these different hypotheses *De Wolf* et al. [11] carried out an electron-microscopic study of the vascular changes in the second trimester when, in normal pregnancy, endovascular trophoblast migrates from the decidual to the myometrial segments of the spiral arteries.

Placental bed biopsies were obtained from 12 hysterectomy specimens between the 14th and 22nd week of gestation. Using optical microscopy to interpret 1-μm thick sections of epon-embedded material, three different stages of vascular modifications can be observed in the myometrial spiral arteries. In stage 1 the vascular wall shows obvious alterations but still contains musculoelastic elements and retains an essentially arterial structure. In stage 2 intraluminal cell plugs are found in close contact with the modified vessel wall. In stage 3 the vessel wall shows characteristic physiological changes with large intramural cells having no contact with the intraluminal cells. These three stages can be seen in a single section through a spiral artery or in different sections through the same spiral artery (fig. 1).

Stage 1. In stage 1 the vessel wall is lined by hypertrophic endothelial cells. The underlying basement membrane is thickened and in some areas is separated from the overlying endothelial cells by large translucent clefts; similar clefts are found focally between the endothelial cells. Myointimal cells infiltrate the intima from the media through fenestrations in the internal elastic lamina. In the intima, between the endothelium and the internal elastic lamina, occasional large stellate cells with the characteristic features of intermediate trophoblast are found and associated with this finding the overlying endothelium is frequently disrupted (fig. 2). In the media there is widening of the intercellular spaces which contain plentiful collagen; some of the smooth muscle cells are hypertrophic and may have prominent basement membranes. These ultrastructural features reflect the light microscopic findings of *Pijnenborg* et al. [22] of swollen endothelium, intimal vacuolation, basophilia of enlarged smooth muscle cells and disorganization of medial tissues. These investigators studied, in hysterectomy specimens with the placenta in situ, the structural alterations in the myometrial spiral arteries from 8 to 18 weeks of gestation applying morphometric and statistical methods to correlate morphological changes in these arteries with the distribution and concentration of migrating interstitial cytotrophoblast. They drew attention to the disruptive changes in the myometrial segments of the placental bed spiral arteries that occur before the time of arrival of endovascular trophoblast, but after the invasion of the myometrium by migrating interstitial trophoblast in the second half of the first trimester. They suggested that non-villous migratory trophoblast acts locally in the placental bed to induce changes in blood vessels to prepare these vessels for subsequent colonisation by endovascular trophoblasts.

Stage 2. The ultrastructural features of the intraluminal cells correspond closely to those of the different morphological types of non-villous trophoblast [21, 25, 32]. At points where the intraluminal cell plugs gain attachment to the intima of the spiral artery there is loss of the maternal endothelium and cells containing the same organelles as the intraluminal trophoblastic cells, but larger, are found in the vessel wall in continuity with the intraluminal cells. In some cases the long narrow branched pro-

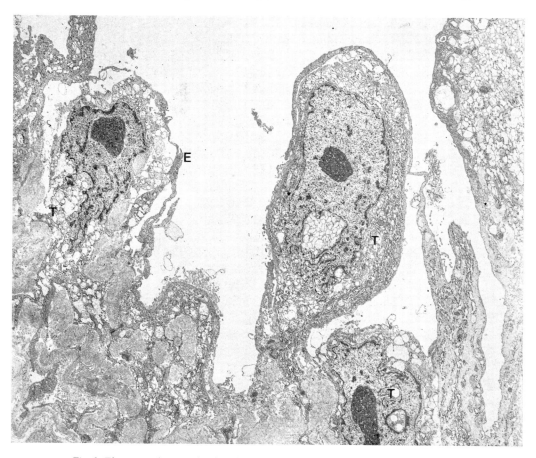

Fig. 2. Electron micrograph of a placental bed spiral artery at the myometrial level at the beginning of the second trimester of normal pregnancy. Stage 1 modification: trophoblastic cells (T) are seen in the intima; the overlying endothelium (E) is disrupted. ×4,600.

cesses of these cells are connected by desmosomes to the protrusions of the stellate cells described in the intima of the vessel wall in stage 1 modifications. The intercellular intimal spaces are large and filled with a mixture of granular and fibrillar material of moderate electron density. Aggregates of fibrils resembling fibrin can also be seen. The disruptive changes in the elastic lamina and the media are clearly more pronounced in stage 2 than in stage 1.

Stage 3. In stage 3 the vessel wall shows the characteristic physiological changes with no direct contact with intraluminal trophoblast. Portions of the vessel wall may be covered by maternal endothelial cell processes or by cell processes from the large intramural trophoblast cells. Occasionally, the extracellular granular and microfibrillar material may be exposed directly to the maternal blood. The remarkable thing with regard to the disruption of the endothelial lining is the infrequency with which this is accompanied by platelet and fibrin deposition, in effect, complicated by thrombosis. A possible explanation for this finding could be the elaboration of some substance by the endovascular and intramural trophoblast which inhibits platelet aggregation [18]. From these findings, we conclude that the three described stages represent three consecutive phases of the trophoblast invasion of the wall of the placental bed spiral arteries. Trophoblast proliferates in the lumen and in the subendothelial space of the spiral arteries. It effectively replaces the normal constituents of the intima of the vessel wall and this is associated with extensive disruption of the internal elastic lamina and subsequent changes in the architecture of the media. Ultimately, the intima is relined by maternal endothelium to complete the physiological changes. The end result is the conversion of small arteries to distended channels of low resistance but high conductance [20, 24] enabling a tenfold increase in blood supply to the intervillous space.

Placental Bed Spiral Arteries in Preeclampsia

Light Optical Microscopy
In 1972, *Brosens* et al. [6] published the important observation that when pregnancy is complicated by preeclampsia or eclampsia, the physiological changes are largely restricted to the decidual segments of the spiral arteries and do not extend to the myometrial segments which retain an essentially normal arterial structure. It is in such vessels, unaffected by physiological changes, that pathological lesions are most easily identified.

In a study of vascular pathology in the hypertensive albuminuric toxaemias of pregnancy, *Hertig* [17] was the first to identify a characteristic arteriopathy which he described as an 'acute degenerative arteriolitis'. His view of its pathogenesis was that 'The lesion in its early stage shows a collection of foamy, fat-laden mononuclear leucocytes or phagocytes beneath the intima of the spiral arteries. This alteration is soon followed by a fibrinoid degeneration of the media, which in turn is superceded by a fibro-

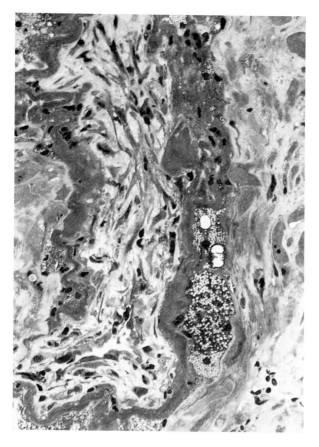

Fig. 3. 1-μm, epon-embedded section through a placental bed spiral artery at the decidual level in a pregnancy complicated by preeclampsia. Necrosis of the media. Fibrin deposition. Myointimal cell proliferation. Advanced lipid accumulation. × 265.

blastic proliferation in the intima resulting in a nearly complete obliteration of the lumen.'

The lesion, now called acute atherosis, is found only in the uterus and only at specific sites, i.e. in those portions of the spiral arteries in the placental bed that are not affected by physiological changes, in the basal arteries, and in the non-placental bed spiral arteries of the decidua vera (fig. 3). There is usually a mononuclear cellular infiltrate around the affected vessel [26]. Those segments of the placental bed spiral arteries affected by physiological changes, having lost their reactive musculoelastic tissue, are relatively immune from acute atherosis [26].

In the absence of physiological changes the spiral arteries in the placental bed are less dilated than in normal pregnancy and the vessel walls may remain responsive to vasomotor influences. There may be further narrowing of the vascular lumen by extensive intimal thickening and by mural and intraluminal thrombosis. These changes could impair blood flow into and through the intervillous space, resulting in placental ischaemia and placental infarction [4] with consequent deleterious effects upon the fetus.

Electron Microscopy of Acute Atherosis

An ultrastructural study of the arteriopathy [10] confirmed that in those pregnancies complicated by preeclampsia, the myometrial segments of the placental bed spiral arteries are not invaded by cytotrophoblastic cells and do not undergo physiological changes; they retain their musculoelastic structure.

The lesion is characterised by focal disruption of the endothelial lining, intimal thickening, fat accumulation in intimal cells and acute degenerative changes in the media. In the intima two cell types can be identified. The first shows the features of smooth muscle cells. The second can be identified as a macrophage type cell. A significant feature seen in many of the intimal cells of both types is a variable degree of cytoplasmic fat accumulation (fig. 4). This ranges from an occasional fat droplet through larger vacuoles, some of which are electron lucent, up to cells containing so many vacuoles as to fill almost the entire cytoplasm reflecting the foamy appearance of these cells at the light microscope level. Cell remnants in the intercellular spaces may show similar vacuoles mixed or in conjunction with other cell organelles.

The intercellular space of the intima is large and contains only a few collagenous and elastic fibres with some fine fibrin aggregates embedded in extensive deposits of flocculent and fine granular material of moderate electron density. In a few places cell debris may be encountered but elsewhere the intercellular space contains non-electrondense material giving it an empty appearance. The expanded extracellular space, partly electron lucent, and the disorganized appearance of the cellular component, may be an expression of oedema and the fibrin aggregates are possibly markers of insudated plasma proteins.

The intima of the myometrial segments of the spiral arteries is demarcated from the media by a fragmented and in places reduplicated elastic

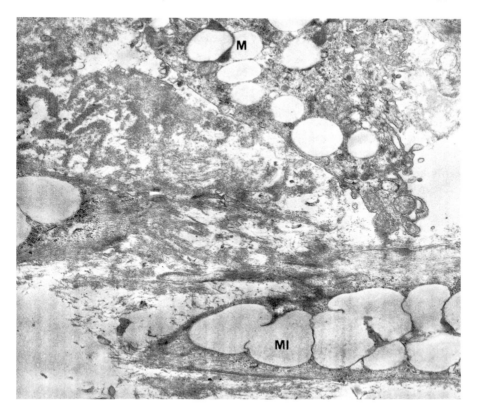

Fig. 4. Electronmicrograph of a placental bed spiral artery at the myometrial decidual junction in a case of preeclampsia. Intimal macrophage (M) and myointimal cells with lipid vacuolation (MI). × 12,200.

lamella. The media, in addition to typical smooth muscle cells, contains cells morphologically identical with myointimal cells. The intercellular space of the media is widened and filled with collagen and elastic fibres, large amounts of follicular and granular material of moderate electron density and many conglomerates of more or less identifiable cell organelles lacking a clear cell membrane. In the intercellular space ghost cells surrounded by fragmented, thick basement membrane substance, but without cell membrane, can be seen.

In summary, the features of acute atherosis of the spiral arteries in severe preeclampsia are: (1) thickening of the intima by oedema, fibrin and other plasma constituents, myointimal cells and macrophages; (2) fatty change in the intimal cells, and (3) necrosis of the media.

Acute atherosis of the spiral arteries has now been described in certain pregnancy disorders other than preeclampsia; in chronic hypertension and in diabetes without hypertension [19], lupus erythematosus [1], idiopathic fetal growth retardation [30] and in women with borderline hypertension [8]. However, it may be that the lesions described by these different authors, although similar, are not identical as there is no agreed definition of the term acute atherosis. The rigid definition proposed by *Robertson* et al. [26] has not been accepted by all other investigators in this subject. The pathogenesis of uteroplacental vasculopathies is probably not the same in these various pregnancy disorders. Analogous lesions have been described in vessels in the kidneys of patients with malignant hypertension, the haemolytic-uraemic syndrome and scleroderma [28] and in rejected renal transplants [7]. The common factor in initiation of these vascular lesions would appear to be endothelial damage, allowing plasma insudation into the intima with fibrin deposition and subsequent migration into the intima of macrophages and myointimal cells. Factors, amongst others, incriminated in causing endothelial damage, are sudden rise in intraluminal pressure, hyperlipidaemia, action of angiotensin, hypoxia and inappropriate immune reactions [12, 27, 31]. The rapid rise in blood pressure and associated vascular spasm are considered to be major factors in the causation of acute atherosis in spiral arteries in preeclampsia. In pregnancy disorders unassociated with hypertension but with vasculopathies other explanations must be sought for the atherotic vascular lesions. Immune complex deposition in artery walls and other immune mechanisms have been postulated but in a recent study, *Kitzmiller* et al. [19] concluded that immunoprotein and complement deposition in decidual arterial lesions in preeclampsia was a consequence rather than a cause of arterial damage. Although there is no good evidence so far to support the concept of inappropriate immune reactions in the pathogenesis of acute atherosis in preeclampsia, this does not exclude the possibility that such mechanisms are operative in other disorders of pregnancy complicated by vascular lesions.

Summary

Ultrastructural study of the placental bed spiral arteries confirms that non-villous cytotrophoblast is involved in the development of the physiological changes occurring in these vessels during normal pregnancy.

The changes observed in the myometrial segments of the spiral arteries before the time of arrival of endovascular trophoblast but after the invasion of the myometrium by migrating interstitial trophoblast, are characterised by widening of the lumen, oedema of the intima, disruption of the elastica and widening of the intercellular spaces of the media. This vascular distension could facilitate the subsequent retrograde migration of endovascular trophoblast. The fetal cells migrate in the vessel lumen and infiltrate the subendothelial space causing further disruption of the arterial intima and media. The altered intima is subsequently recovered by the endothelium. In this way, the cytotrophoblast is incorporated into the wall of the placental bed spiral arteries which are converted from small muscular arteries into distended hyalinized tubes.

In pregnancies complicated by preeclampsia and in some pregnancies complicated by fetal growth retardation, these physiological changes are largely restricted to the decidual segments leaving the myometrial segments unaffected. The lesion of acute atherosis is characterised by thickening of the intima and necrosis of the media. The intimal thickening is due to deposition of fibrin and other plasma constituents and migration into the intima of macrophages and myointimal cells which accumulate fat in their cytoplasm to become foam cells. Clinical and experimental studies indicate that these lesions can be initiated by several factors which cause endothelial injury.

References

1 Abramowsky, C.R.; Vegas, M.E.; Swinehert, G.; Gyves, M.T.: Decidual vasculopathy of the placenta in lupus erythematosus. New Engl. J. Med. *303:* 668–672 (1980).
2 Adamson, K.; Meyers, R.E.: Circulation in the intervillous space; obstetrical considerations in fetal deprivation; in Gruenwald, The placenta and its maternal supply line, pp. 158–177 (University Park Press, Baltimore 1975).
3 Boyd, J.D.; Hamilton, W.J.: Development and structure of the human placenta from the end of the 3rd month of gestation. J. Obstet. Gynaec. Br. Commonw. *74:* 161–166 (1967).
4 Brosens, I.; Renaer, M.: On the pathogenesis of placental infarcts in preeclampsia. J. Obstet. Gynaec. Br. Commonw. *79:* 794–799 (1972).
5 Brosens, I.; Robertson, W.B.; Dixon, H.G.: The physiological response of the vessels of the placental bed to normal pregnancy. J. Path. Bact. *93:* 569–579 (1967).
6 Brosens, I.; Robertson, W.B.; Dixon, H.G.: The role of the spiral arteries in the pathogenesis of preeclampsia; in Wynn, Obstetrics and gynecology annual, pp. 177–191 (Appleton Century Crofts, New York 1972).
7 Dempster, W.J.; Harrison, C.V.; Shackman, R.: Rejection processes in human homotransplanted kidneys. Br. med. J. *ii:* 969–976 (1964).
8 De Wolf, F.; Brosens, I.; Renaer, M.: Fetal growth retardation and the maternal arterial supply of the human placenta in the absence of sustained hypertension. Br. J. Obstet. Gynaec. *87:* 678–685 (1980).
9 De Wolf, F.; De Wolf-Peeters, C.; Brosens, I.: Ultrastructure of the spiral arteries in the human placental bed at the end of normal pregnancy. Am. J. Obstet. Gynec. *117:* 833–848 (1973).

10 De Wolf, F.; Robertson, W.B.; Brosens, I.: The ultrastructure of acute atherosis in hypertensive pregnancy. Am. J. Obstet. Gynec. *123:* 164–174 (1975).
11 De Wolf, F.; De Wolf-Peeters, C.; Brosens, I.; Robertson, W.B.: The human placental bed. Electronmicroscopic study of trophoblastic invasion of the spiral arteries. Am. J. Obstet. Gynec. *137:* 58–70 (1980).
12 Gerrity, R.G.: Transition of blood-borne monocytes into foam cells in fatty lesions. Am. J. Path. *103:* 181–190 (1981).
13 Gruenwald, P.: The supply line of the fetus: Definitions relating to fetal growth; in Gruenwald, The placenta and its maternal supply line, pp. 1–17 (University Park Press, Baltimore 1975).
14 Hamilton, W.J.; Boyd, J.D.: Development of the human placenta in the first three months of gestation. J. Anat. *94:* 297–328 (1960).
15 Hamilton, W.J.; Boyd, J.D.: Trophoblast in human utero-placental arteries. Nature, Lond. *212:* 906–908 (1966).
16 Harris, J.W.S.; Ramsey, E.H.: The morphology of human uteroplacental vasculature. Contr. Embryol. *38:* 43–58 (1966).
17 Hertig, A.T.: Vascular pathology in the hypertensive albuminuric toxemias of pregnancy. Clinics *4:* 602–613 (1945).
18 Hutton, R.A.; Chow, F.P.R.; Cnaft, I.L.; Dandona, P.: Inhibitors of platelet aggregation in the fetoplacental unit and myometrium with particular reference to the ADP-degrading property of placenta. Placenta *1:* 125–130 (1980).
19 Kitzmiller, J.L.; Watt, N.; Driscoll, S.G.: Decidual arteriopathy in hypertension and diabetes in pregnancy. Immunofluorescent studies. Am. J. Obstet. Gynec. *141:* 773–779 (1981).
20 Moll, W.; Künzel, W.; Herberger, J.: Hemodynamic implications of hemochorial placentation. Eur. J. Obstet. Gynec. reprod. Biol. *5:* 67–74 (1975).
21 Okudaira, Y.; Suzuki, S.; Okudaira, M.; Hashimoto, T.; Hayakawa, K.: Electron microscopic observations on the formation of syncytiotrophoblast from cytotrophoblast. J. Elec. Microsc. *17:* 47–54 (1968).
22 Pijnenborg, R.; Bland, J.M.; Robertson, W.B.; Brosens, I.: Uteroplacental arterial changes related to interstitial trophoblast migration in early human pregnancy. Placenta (submitted for publication).
23 Pijnenborg, R.; Dixon, G.; Robertson, W.B.; Brosens, I.: Trophoblastic invasion of human decidua from 8 to 18 weeks of pregnancy. Placenta *1:* 3–19 (1980).
24 Ramsey, E.M.; Chez, R.A.; Doppman, J.: Radioangiographic measurement of the internal diameters of the uteroplacental arteries in rhesus monkeys. Am. J. Obstet. Gynec. *135:* 247–251 (1979).
25 Robertson, W.B.; Warner, B.: The ultrastructure of the human placental bed. J. Obstet. Gynec. *112:* 203–211 (1974).
26 Robertson, W.B.; Brosens, I.; Dixon, H.G.: Maternal uterine vascular lesions in the hypertensive complications of pregnancy; in Lindheimer, Katz, Zuspan, Hypertension in pregnancy, pp. 115–127 (Wiley, New York 1976).
27 Ross, R.; Clomset, J.A.: The pathogenesis of atherosclerosis. New Engl. J. Med. *295:* 369–377, 420–425 (1976).
28 Sinclair, R.A.; Antonovych, T.T.; Mostofi, F.K.: Renal proliferative arteriopathies and associated glomerular changes. A light and electron microscopic study. Human Pathol. *7:* 565–588 (1976).

29 Sheppard, B.L.; Bonnar, J.: The ultrastructure of the arterial supply of the human placenta in early and late pregnancy. J. Obstet. Gynaec. Br. Commonw. *81:* 497–511 (1974).
30 Sheppard, B.L.; Bonnar, J.: The ultrastructure of the arterial supply of the human placenta in pregnancy complicated by fetal growth retardation. Br. J. Obstet. Gynaec. *83:* 948–959 (1976).
31 Still, W.J.S.: Intima of the hypertensive rat. Archs Path. *89:* 392–404 (1970).
32 Wynn, R.M.: Cytotrophoblastic specialisations. An ultrastructural study of the human placenta. Am. J. Obstet. Gynec. *114:* 339–355 (1972).

F. De Wolf, MD, Academisch Ziekenhuis St. Rafael, Verloskunde, Gynaecologie, Kapucijnenvoer 33, B-3000 Leuven (Belgium)

Placental Morphology of Low-Birth-Weight Infants Born at Term

Gross and Microscopic Study of 50 Cases

A.G.P. Garcia[1]

Instituto Fernandes Figueira, Fiocruz, Rio de Janeiro, Brasil

Introduction

Despite the existence of many clinical studies on newborn infants who are small with respect to their gestational age, the associated placentas have rarely been studied.

The purpose of the present work is to investigate the appearance of the placentas of 50 low-birth-weight infants born at term in Rio de Janeiro, Brasil.

Material and Methods

Routine, gross examination of all placentas used in this study (obtained from the delivery rooms of the Maternidade Clovis Corrêa da Costa, Instituto Fernandes Figueira, Rio de Janeiro, Brasil) was performed using *Benirschke and Driscoll's* [3] technique.

The macroscopic examination was preceded by the evaluation of the prenatal and natal data and of the postnatal history. For the microscopic examination six blocks of placental tissue were chosen; one from the margin, another from near the center (making sure that the latter contained a representative sample of chorionic vessels), and two or three blocks from abnormal areas. A roll of the membranes, a longitudinal section and a cross-section of the umbilical cord were examined. Formalin-fixed blocks were embedded in paraffin and sections stained with hematoxylin-eosin. Special stainings were used in the elucidation of the lesions: Perl's method to identify iron pigment; Kossa, for calcium deposits; Shorr, Giemsa and periodic acid-Schiff for the detection of viral inclusions or

---

[1] With the collaboration of: *Hilda I.B. Ramos*, MD, *Maria Evangelina F. Fonseca*, PharmD, *Heloisa N. Outani*, MD, *Marisa A. Ferreira*, MD, *Marcus M. Jesus*, MD, *Paulo Geraldo Silva*, MD, *Rosane R. Mello*, MD.

Table I. Placental lesions in 50 low-birth-weight infants born at term

Primary diagnosis[1]	Number	Percent
Placental infections	37	74
Villitis, unknown	28	
Villitis, cytomegalovirus	3	
Villitis, T. gondii	2	
Villitis, rubella virus	1	
Villitis, enterovirus	1	
Acute chorioamnionitis and funiculitis	2	
Placental 'circulatory disturbances'	7	14
Gross anomalies	3	6
Chorioangiomatosis	1	2
Villous dismaturity	2	4
Total	50	100

[1] Many placentas exhibited more than one pathologic feature. The classification was made according to the lesions considered to be of primary significance.

micro-organisms; Gram's staining in the study of bacterial morphology and Levaditi's or Wharthin-Starry's method for the identification of *Treponema pallidum*.

Complete gross and light-microscopic studies were performed on 50 placentas of infants born at term (38 weeks or more), whose birth weight was 2,500 g or less. The data obtained constitute the basis of the present paper.

Classification

The placentas were evaluated according to the criteria of *Altshuler* et al. [1]. Table I summarizes the placental lesions detected in our 50 cases.

Results

Villitis

Of the 50 specimens, 37 (74%) exhibited gross and microscopic lesions compatible with a diagnosis of hematogenous infection. It was based on the presence of villitis. This was characterized by intrinsic inflammatory response occurring within the villi, due to the proliferation and infiltration of fetal cells. The villitides were focal or diffuse, and developed at random throughout the villous plate, consisting of single or multiple neighboring villi. They occurred in free villi and commonly in villi that were attached to the decidua, assuming various histological appearances [2]: necrotizing, proliferative, reparative and stromal fibrosis. In the great majority of cases,

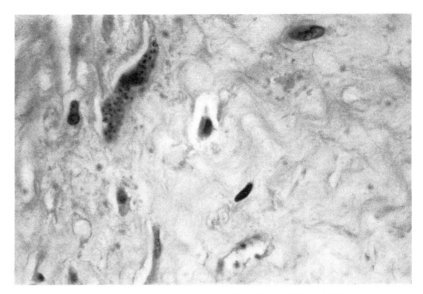

Fig. 1. Umbilical cord. Toxoplasma cyst in a fibroblast. HE. ×1,230.

vascular involvement was observed, including endovascular or concentric sclerosis of the vessel wall with luminal occlusion. It is noteworthy that the vascular involvement included the umbilical and chorionic vessels, as well as the fetal stem vessels, the villous capillaries and the decidual ones.

In 2 cases we could establish the etiology of the infectious process due to the presence of trophozoites and cystic forms of *Toxoplasma gondii* (fig. 1). In 1 of these cases, the serological diagnosis could be made (specific IgM) and confirmed the morphologic one. This type of villitis showed some peculiarities: necrobiosis of the trophoblast cells, subepithelial histiocyte infiltrate (fig. 2), consistent finding of vasculitis and fibrosis of the stroma. Focal or nodular villitis and intervillitis with and without cystic forms of the parasite were occasionally seen.

In 1 case, besides the laboratory diagnosis of maternal rubella during pregnancy, microscopic examination of the placenta pointed out the presence of proliferative-necrotic villitis with selective involvement of the endothelium of villous capillaries and stromal fibrosis. In the decidual cells, trophoblastic epithelium, capillary endothelium and in the stroma cells of the villi, nuclear and protoplasmatic round, eosinophilic inclusions were observed (fig. 3).

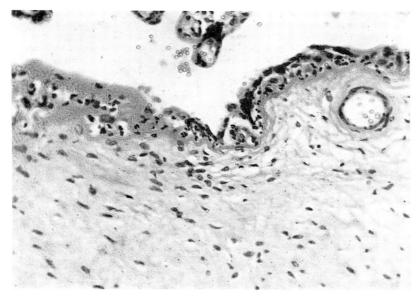

Fig. 2. Surface of a chorionic stem with necrosis of the trophoblast epithelium and presence of subepithelial cellular infiltrate. HE. ×125.

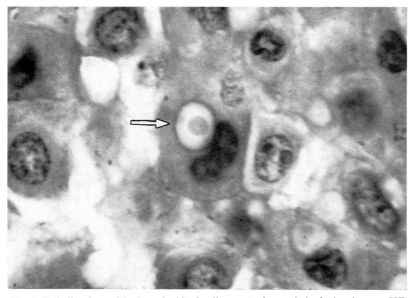

Fig. 3. Rubella placentitis. In a decidual cell a protoplasmatic inclusion is seen. HE. ×1,250.

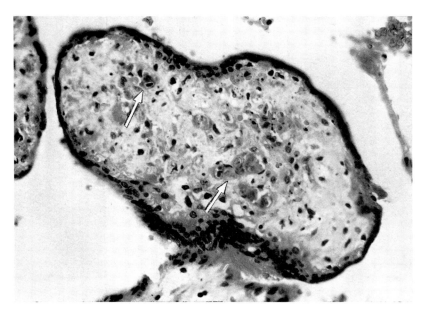

Fig. 4. Placental villus. Presence of groups of cells in cytomegalic degeneration. HE. ×510.

In 3 cases, cytomegalovirus infection was strongly suggested by the morphological aspect of some of the villi. Focal necrotizing, proliferative or evanescent villi were observed, as well as foci of villous stromal fibrosis. A significant finding was that of severely compromised villi, with partial necrosis of the trophoblast, where some degenerated cells exhibited an appearance suggestive of cytomegalic cells (fig. 4). Frequently, the presence of large quantities of hemosiderin pigment was observed, impeding the complete individualization of the cellular structure. In 1 of these 3 cases, serologic examination could be made and demonstrated the presence of antibodies to cytomegalovirus (1/32) in the blood of the newborn.

An enterovirus (Echovirus type 11) was isolated in 1 case. There was severe and disseminated proliferative-necrotic villitis with selective involvement of the fetal vascular circuit (fig. 5).

In 28 of 37 cases (76%) where villitis was clearly identified, the etiology of the infection could not be determined by morphologic examination and it was not possible to make complementary investigations.

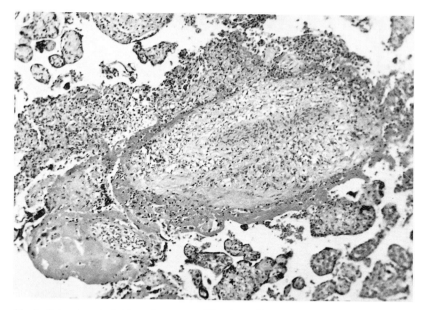

Fig. 5. Echovirus infection. Severe proliferative villitis with endovascular proliferation. HE. ×115.

Placental Circulatory Disturbances

In this group, 7 specimens were included, 3 associated with clinical histories of toxemia during pregnancy. The weight of the placentas was severely reduced, or near the minimum normal weight; the placental to fetal weight ratio was between 1/5 and 1/8,5 (normal 1/5–1/7). The frequency of infarction and retrodecidual hematomas was higher than in the other groups. The most characteristic microscopic feature was the rarefaction of free villi, which exhibited a peculiar appearance.

The terminal villi were hypovascular with small nondilated vessels, associated with obliterative endarteritis of the fetal stem arteries.

The hypovascular villi exhibited a large number of syncytial knots (fig. 6), no vasculo-syncytial membrane, cytotrophoblastic hyperplasia, thickening of the trophoblastic basement membrane and an increased content of stromal collagen. It is noteworthy that fibrin was deposited between and within the villi with proliferation of intervillous cytotrophoblasts. Decidual arteriopathy and consequent ischemic lesions were always present.

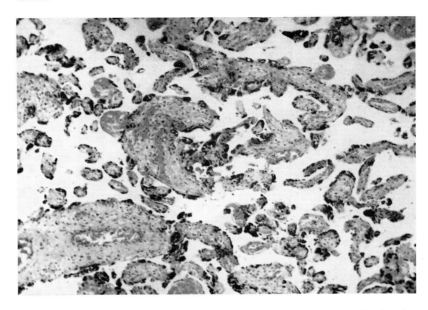

Fig. 6. Maternal hypertension: a panoramic view of the villous plate showing dystrophic villi and abundance of syncytial knots. HE. ×44.

It manifested by fibrinoid necrosis of the arterial wall allied to intramural accumulation of lipid-laden macrophages and perivascular lymphocytic cuffing. In 1 of the cases of toxemia, proliferative villitis of unknown etiology was also detected.

Macroscopic Abnormalities of the Placenta

Three placentas were classified in this group. Abnormal insertion of the umbilical cord (velamentous or pedunculated) and 1 case of placenta circumvallata were observed (fig. 7, 8). In other cases, even though gross abnormalities were detected, other causes seemed more expressive in the clinicopathologic evaluation.

A case of diffuse chorioangiomatosis was registered (fig. 9). Though it was related to the presence of focal villitis, fetal growth retardation was attributed to the vascular villous lesion.

Villous dismaturity and avascular villi associated with obliteration of the vessels in the villous stems was considered the only expressive factor in 2 placentas.

Placental Morphology of Low-Birth-Weight Infants Born at Term

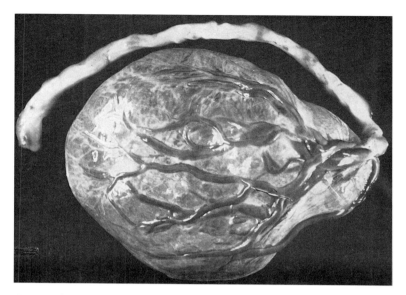

Fig. 7. A battledore placenta with velamentous insertion of the umbilical cord.

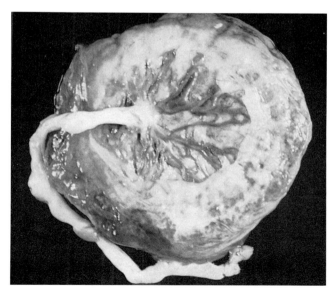

Fig. 8. Placenta extrachorales, partially circumvallata.

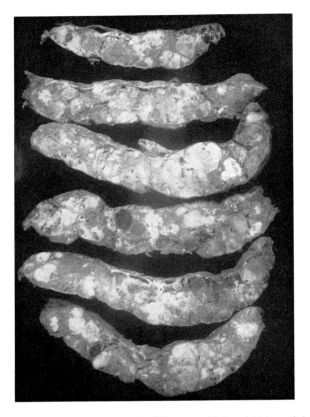

Fig. 9. Chorioangiomatosis diffusa: multiple whitish nodules in the villous plate could be observed.

Maternal anemia was verified in 4 cases, 3 of them characterized as erythrofalcemia. The diagnosis was made by the morphology of maternal red cells in the intervillous space (fig. 10); in none of them were maternal clinical abnormalities detected.

Comments

Despite the existence of innumerable multidisciplinary studies about low-birth-weight infants, the pathogenesis of intrauterine growth retardation is still not clearly understood. However, there appears to be a consensus

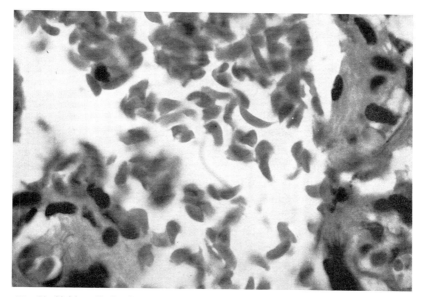

Fig. 10. Sickle cells in the intervillous space. HE. ×1,250.

that the associated factors are related intrinsically to the fetus, to the fetoplacental unit and to the maternal organism. *Gruenwald* [5] stated that placental lesions are the result of the same adverse factors which produce small-for-date infants. *Grundmann* [6], although considering that a primary placental cause is not common, emphasizes that the study of the placental lesions will indirectly reveal conditions that perhaps can be cured and eventually prevent fetal wastage.

Our data seem to corroborate the findings of *Altshuler* et al. [1] who pointed to placental abnormalities of causal significance in 58 of 63 placentas of small-for-date infants. In their series, approximately 25% of the placentas showed villitis of unknown etiology, probably viral. In the present study, 37 of 50 placentas (74%) presented lesions compatible with hematogenous infection. Only in 7 cases was the etiology of the intrauterine infection detected: cytomegalovirus – 3; rubella virus – 1; type 11 Echovirus – 1; *T. gondii* – 2. This indicates the need of complementary laboratory examination of the placenta (virologic, bacteriologic, etc.) for diagnosis and prevention of these infections. Bacterial chorioamnionitis and funisitis (2 cases), due to their acute nature, were not considered as causes of intrauterine growth retardation.

Grundmann [6] considers villitis an important cause of intrauterine growth retardation, more frequent in communities of low socioeconomic status, as in our country. It is worth noting that in 50 pregnant women included in this study a high degree of maternal malnutrition was not observed.

In 14% of our cases, placental abnormalities related to circulatory disturbances associated with maternal hypertension were observed. Some of the aforementioned villous abnormalities are attributed to ischemia, which is a direct consequence of reduced maternal blood flow related to hypertensive changes in the decidual vessels.

Sheppard and Bonnar's [10] experience, based on biopsies of the placentary bed, indicates that, although hypertension will intensify the response of decidual vessels, similar lesions may be found in pregnancies with low-birth-weight fetuses without hypertension. They consider that the lesions in the uteroplacental vasculature, which appear to be a feature of fetal growth retardation, are related more to an abnormal maternal response to trophoblast than to hypertension.

It is considered that extensive placental infarction is the hallmark of a grossly abnormal vascular tree and of severely compromised maternal uteroplacental circulation. It is these factors that are the true cause of the apparent effects of placental infarction on fetal growth rather than the simple destruction of villous tissue.

Fox [4] emphasizes that it seems unreasonable to postulate that the apparent ill effects of maternal hypertension on the fetus are due to placental insufficiency rather than to the inability of the mother to supply the fetus with an adequate amount of oxygen. The placenta, far from functioning at the optimal level which the maternal condition will allow, often shows changes of a compensatory nature which are designed to diminish the effects of the failure of maternal circulation.

Shanklin [9], examining 6,500 placentas, 10% associated with low-birth-weight in term infants, identified in this group major grossly observable anomalies attributed mainly to abnormal fetal-placental dynamics. A large number of placentas exhibiting abnormal configurations, infarction and widespread subchorial thrombosis were present in *Grundmann's* [6] series. Using angiography and scanning electron microscopy, he observed an abnormal type of cotyledon with an abnormal vascular pattern, as well as a predominance of immature villi. In extrachorial placentas and a single umbilical artery a small number of placental lobes were detected. He concluded that abnormal arterial supply was the cause of these placental lesions.

In 3 cases of the placentas we studied, gross abnormalities were considered the main cause of the retarded intrauterine fetal growth. Noteworthy is a case of diffuse chorioangiomatosis, which seemed to be the cause of the fetal dismaturity, although focal villitis had also been observed.

In most of the specimens analyzed, villous dismaturity and avascular terminal villi were observed, and were the only abnormalities found in 2 placentas. One of the mothers was a heavy smoker, and it is known that chronic inhalation of nicotine [7, 11] during pregnancy will result in decreased growth of the fetus due to subnormal perfusion of the intervillous space.

Finally, we must comment on maternal anemia, the most abnormal composition of maternal blood in pregnancy [8]. It is commonly associated with smaller fetuses having relatively large placentas, as it is known that some hemoglobinopathies interfere with fetal growth. In the sicklemia, besides a reduced concentration of hemoglobin in the maternal blood, there is increased viscosity of the blood, mainly during the phases of sickling, both contributing to the reduction of the oxygen supply of the fetus.

In 2 of the 4 reported cases of maternal anemia, there was an association with toxemia; in the others, a diffuse villitis was evident. We considered that maternal anemia was not the main cause of retardation of the fetal growth, but acted as a contributory cause.

Summary

Gross and microscopic examination in 50 placentas of low-birth-weight infants showed: (1) hematogenous infection (74%); (2) placental circulatory disturbances related to maternal hypertension (14%); (3) abnormal placentation (6%); (4) isolated villous dismaturity (4%), and (5) diffuse chorioangiomatosis (2%).

As the main placental lesion associated to low-birth-weight in this series was hematogenous infection, the author stress the validity of the virologic, bacteriologic and parasitologic examination of the placenta combined with the morphologic one in the detection of the etiology of intrauterine infection.

References

1 Altshuler, G.; Russel, P.; Ermocilla, R.: The placental pathology of small for gestational age infants. Am. J. Gynec. *121:* 351–359 (1975).
2 Altshuler, G.; Russel, P.: The human placental villitides. A review of chronic intrauterine infection. Curr. Top. Pathol. *60:* 64–112 (1975).

3 Benirschke, K.; Driscoll, S.G.: The pathology of the human placenta, pp. 4–9 (Springer, New York 1967).
4 Fox, H.: The placenta in maternal disorders; in Bennington, Pathology of the placenta, vol. VII. Major problems in pathology, pp. 213–237 (W.B. Saunders Company Ltd., London 1978).
5 Gruenwald, P.: The relation of deprivation to perinatal pathology and late sequels; in The placenta and its maternal supply line, pp. 335–356 (University Park Press, Baltimore 1975).
7 Grundmann, E.: Perinatal pathology. Curr. Top. Pathol. *66:* 2–55 (1979).
8 Mulcary, R.; Murphy, J.; Martin, F.: Placental changes and maternal weight in smoking and nonsmoking mothers. Am. J. Obstet. Gynec. *106:* 703–704 (1970).
Singla, P.N.; Chand, S.; Soshilla, K.; Agaewal, K.N.: Effect of maternal anemia on the placenta and the newborn infant. Acta paediat. scand. *67:* 645–648 (1978).
9 Shanklin, D.R.: The influence of placental lesions on the newborn infant. Pediat. Clins N. Am. *17:* 25–42 (1970).
10 Sheppard, B.L.; Bonnar, J.: An ultrastructural study of utero-placental spiral arteries in hypertensive and normotensive pregnancy and fetal growth retardation. Br. J. Obstet. Gynaec. *88:* 695–705 (1981).
11 Wingerd, J.; Christianson, R.; Schoen, E.J.: Placental ratio in white and black women: relation to smoking and anemia. Am. J. Obstet. Gynec. *124:* 671–675 (1976).

A.G.P. Garcia, MD, Instituto Fernandes Figueira, Avenida Ruy Barboza, 716, Rio de Janeiro Cap. 22.250 (Brasil)

Placentitis

Geoffrey Altshuler

Departments of Pathology and Pediatrics and The Oklahoma Children's Memorial Hospital, The University of Oklahoma Health Sciences Center, Oklahoma City, Okla., USA

Introduction

The placenta can be a valuable specimen if it is used promptly and clinically like all other surgical specimens. Placental immunopathologic and ultrastructural studies are occasionally helpful. Usually, however, routine hematoxylin and eosin light-microscopic sections are sufficient to initiate or support precise diagnoses. Characteristic patterns of placentitis are associated with infectious agents of congenital disease. From histopathology, therefore, it may be possible to adduce dual infection or placentitis of more than one cause. Chronic fetal infectious disease, previously considered to result from transplacental infection, often results from ascending intrauterine infection.

Routes of Infection

Benirschke [17] and *Blanc* [25] separately pioneered understanding of means by which a fetus may become infected. Within the past few years, this knowledge of placental histopathology has been greatly augmented by contributions of virologists and microbiologists. A diversity of viruses, anaerobic bacteria, mycoplasma and chlamydia may infect the mother and fetus and be responsible for overt or latent congenital infection.

Severe placental sinusoidal leukoerythrostasis is histopathologic evidence of bacteremia, viremia or endotoxemia.

All fetal infections are derived 'transplacentally' either by blood-borne infection via the mother's placental sinusoids or by direct ascent of organ-

Fig. 1. Chemotaxis in a placental surface vessel. Note the severe presence of inflammatory cells exclusively on the side of the vessel adjacent to the amniotic surface. HE. × 80.

isms in the vagina and cervix [26]. Ascending infection spreads from the membranes to the placenta and thereby causes chorioamnionitis. This placentitis is augmented by chemotactic infectious agents within the amniotic fluid. In chorioamnionitis, amniotropism of placental surface inflammatory cells is evidence of an infectious etiology (fig. 1). Hypoxia and acidosis, for example, do not produce amniotropism. Amniotic fluid has been shown to include antibacterial factors [28, 35, 79]. In African populations, nutritional zinc deficiency may account for an increased incidence of chorioamnionitis [92].

When organisms ascend through placental membranes (membranitis), a fetus may become infected in the absence of elevated maternal serology.

If an infectious diseases laboratory is not advised of possible infection by *Listeria monocytogenes*, anaerobic organisms, chlamydia, mycoplasma and viruses, it is likely that these infections will not be substantiated by culture isolation.

Ascending Intrauterine Infection

Chorioamnionitis is so very common in placentas of premature newborns that it is difficult to determine placental signs of neonatal pneumonia and sepsis. In a study of placental pathology compared with clinical outcome, chorionic microabscesses and vasculitis in all three vessels of the umbilical cord were significantly associated with neonatal sepsis [44].

Newborns of less than 32 weeks gestation often have chorioamnionitis that is unassociated with umbilical cord vasculitis and funisitis [19]. For these newborns, frozen section examination of placental tissue should be performed additional to light-microscopic evaluation of the umbilical cord.

In my experience, the incidence of deciduitis varies strikingly with gestational age: 20 weeks gestation – 100%; 30 weeks gestation – 40%; 36 weeks gestation – 25%; 40 weeks gestation – 10%. *Russell* [74] has reported a similarly high incidence in a large population studied in Sydney, Australia.

Obstetric factors that correlate with chorioamnionitis include membranes that are ruptured for more than 24 h, foul smelling amniotic fluid, maternal fever, tachycardia and leukocytosis, fetal tachycardia and the presence of a cervical ligature [36, 74].

The Importance of Chorioamnionitis

Naeye and Blanc [57] have prolifically reported the association of chorioamnionitis with fetal morbidity and mortality. Chorioamnionitis is frequently associated with congenital pneumonia, even when no pathogens are isolated [17, 25]. In a study of 783 consecutive neonatal autopsies, 36% had congenital pneumonia and/or chorioamnionitis [57]. This amniotic fluid infection syndrome is also especially associated with recurrent spontaneous abortion and with prematurity-related morbidity and mortality; it occurs in almost 55% of mothers who suffer two or more pregnancy losses [57].

Table I. Incidence of maternal and chronic perinatal infection

Infection	Mother (per 1,000 pregnancies)	Fetus[1] (per 1,000 livebirths)	Newborn[2] (per 1,000 livebirths)
CMV	40–150	5–15	50–100
Rubella			
Epidemic	20–40	4–30	0.0
Interepidemic	1	0.5[3]	0.0
Toxoplasmosis	1.5–6.4	0.75–1.3	0.0
HVH	10–15	rare	0.03 to 0.33
Syphilis	0.2[4]	0.1[4]	0.0
Cumulative total (excluding epidemic rubella)	53–173	6.3–22	50–100

[1] Designates intrauterine infection acquired prior to the immediate delivery period.
[2] Designates infection acquired at or just prior to delivery.
[3] Estimate.
[4] Estimate derived from Center for Disease Control surveillance data.

Data of the United States Collaborative Perinatal Project establishes that the amniotic fluid infection syndrome is the leading cause of perinatal mortality, 17% of deaths being associated with this condition [59]. The syndrome assumes even greater significance because it exacerbates effects of hyperbilirubinemia in the production of kernicterus and psychomotor retardation [60].

Coitus in the last month of pregnancy may cause chorioamnionitis and fetal morbidity and mortality [61]. A study from *Mills* et al. [52] in Israel has provided data which challenge this conclusion. Until such time as the issue is resolved by clarification of biostatistical differences in study design, *Herbst's* [42] opinion may be the most advisable: 'A reasonable policy might be to recommend the avoidance of intercourse and orgasm in the third trimester in women with a poor reproductive history or in those who, on pelvic examination, have premature ripening of the cervix.'

Amniotic fluid infections occur two to three times more often when membranes rupture just before labor, in comparison with just after labor [62]. Thus, chorioamnionitis precedes premature rupture of membranes more often than it results from that entity [22, 58, 62].

Etiologic Causes of Ascending and Hematogenous Infection

Approximately 60% of neonatal sepsis and meningitis is attributable to group B β-hemolytic streptococcus and *Escherichia coli; Staphylococcus aureus, Klebsiella enterobacter*, group D *Streptococci* and *Pseudomonas aeruginosa* are common causes of neonatal sepsis but *Streptococcus pneumoniae, Neisseria meningitidis* and *Hemophilus influenzae* are rare [82]. Anaerobic organisms that cause ascending congenital infection include *Peptococcus, Peptostreptococcus, Fusobacterium, Lactobacillus* and *Bacteroides* such as *Bacteroides fragilis* and *Bacteroides vulgaris*. Although anaerobic organisms have been isolated from blood cultures of 26% infants studied, anaerobic bacteremia is usually self-limited and is rarely life-threatening [82].

Toxoplasmosis, rubella virus, cytomegalovirus (CMV), herpes simplex virus and syphilis produce 'Torch' infections, with growth retardation, hepatosplenomegaly, purpura and intracranial calcifications [65]. Of these, CMV disease is by far the most common [2] (table I).

Multiple Chronic Congenital Infections

Infection with two or more venereally acquired organisms is probably significantly more common than thus far reported [14, 78]. Placental examinations can be diagnostic but are often lacking (fig. 2). Alleged co-infections of CMV and toxoplasma have not always been adequately documented [23].

Selected Considerations of Placentitis and Congenital Infection

CMV

Most infants who excrete CMV have acquired their infection beyond 3 months age [87]. CMV excretion at birth occurs in more than 2% of all newborns in England and Finland, but has not been reported in Japan, Thailand and Guatemala [87]. About 50% of middle-class women enter the childbearing age with serologic evidence of previous CMV infection [70]. As many as 28% of pregnant women excrete CMV from the cervix, maximally so in third-trimester oriental populations [1, 51, 53, 67].

The greater majority of CMV-infected fetuses acquire disease from mothers who experience recurrent CMV infection and only in a minority of primary maternal infection does symptomatic neonatal CMV disease

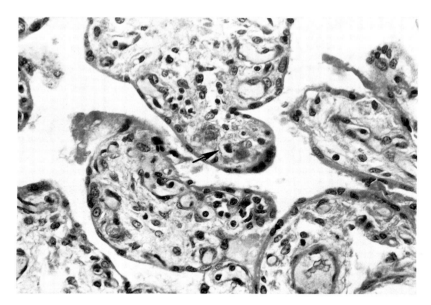

Fig. 2. Dual infection. An intranuclear virus inclusion is present at the center of this illustration (arrow). Two adjacent large cells contain numerous cytoplasmic inclusions. The associated newborn was additionally proven, by clinical and pathologic studies, to suffer from syphilis. HE. × 500.

occur [72]. Postnatal acquisition of CMV occurs in 58% of infants fed CMV-infected breast milk and as many as 13% of mothers excrete virus by this means as opposed to 2.4% from the throat, 4.7% from the urine and 10.1% from the cervix [41, 87].

CMV is venereally transmitted. It resides in semen, the cervix and endometrial tissue [29, 46, 96], and is therefore capable of producing ascending intrauterine infection. Neonatal pneumonitis may be acquired by this pathogenesis [15, 97].

Waner and I have produced relevant findings in an in vivo animal model. We inoculated virus into the amniotic sacs of mouse fetuses, at 17–18 days gestation, and obtained inflammatory cell infiltrates and numerous CMV-infected yolk sac epithelial cells.

To date, human CMV placentitis has been considered characteristic of hematogenous infection. CMV lymphoplasmacytic villitis in placentas of first-trimester fetuses has been reported to be morphologic evidence of immunologic competence [6, 26]. Review of the literature shows that CMV

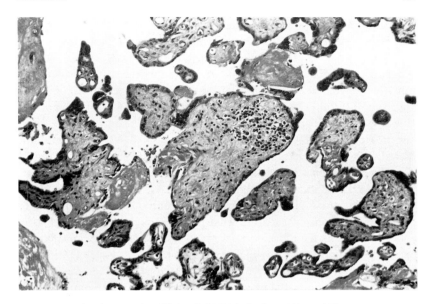

Fig. 3. Lymphoplasmacytic villitis of CMV infection. HE. ×200.

placentitis is typically manifested by focal lymphocytic and/or plasmacytic villitis (fig. 3), focal nonspecific proliferative villitis, necrotizing fetal vasculitis with thrombotic changes, avascular villi, focal hemosiderin deposits and, often, villous edema. Pathognomonic virus inclusions are readily appreciable in less than 50% of placentas from symptomatic newborns [8, 23, 26]. Granulomatous inflammation is not a feature of CMV placentitis [8, 23, 26]. Its presence in the placenta of a first-trimester fetus may therefore have resulted from dual infection [8, 23, 26].

Rapid ultrastructural diagnosis of congenital CMV infection is easily achievable within 1 h [48, 55]. This is obtained in 92% of cases wherein congenital CMV is diagnosed by electron-microscopic examination of urine [86].

Herpes Simplex Virus

Follow-up information of mothers whose infants have succumbed to herpes simplex virus infection is very much lacking. An observation was recently poignantly made by a father whose first child succumbed in infancy to herpes encephalitis: 'A review of the literature failed to reveal any documentation of normal offspring (of afflicted mothers) in successive pregnancies... I appeal... for any information in this regard' [49].

Symptomatic herpetic infection in the newborn has been estimated to occur in 1 of 7,500 deliveries [63]. Of women whose newborns suffer symptomatic congenital infection, 21% have experienced prior spontaneous abortion [66, 98]. Cervical herpes simplex infection is well known [10]. Histologically recognized endometrial herpetic inclusions have been reported but, because it is difficult to identify them by light-microscopic examination, their incidence in uterine curettings and placental tissue is uncertain [37].

Herpes simplex can produce ascending infection by viral ascent from the cervix to the uterus and placenta [5, 26, 37, 39, 64].

The placental histopathology of congenital herpes simplex infection has only occasionally been reported [26]. Villous necrosis and plasma cell chorioamnionitis have been observed [5, 100]. Recently, *Benirschke* [24] has encountered fatal second-trimester fetal infection associated with lymphohistiocytic chorioamnionitis but unassociated with villous necrosis.

Fetal infection with herpes simplex virus has been reported to cause skin lesions, microcephaly, diffuse brain damage and chorioretinitis [32, 53, 83, 89]. Although herpes viruses are thus potentially teratogenic, a numerically significant causal relation between herpes virus infection and fetal malformation has not been established [47]. The neonatal effects of congenital herpes simplex infection may be so variable as to be clinically inapparent [11, 88, 93, 95]. Based on clinical experience of 58 cases, recommendations for the management of herpes simplex virus infection during pregnancy have been recently provided [38].

Toxoplasmosis

The incidence of symptomatic congenital toxoplasmosis is considerably less than that of CMV and herpes simplex. Nevertheless, it has been carefully estimated that the cost of caring for congenitally infected children born each year in the US is $ 221.9 million [99]!

Six of every 1,000 US women acquire toxoplasmosis during pregnancy [81]. Approximately half of these mothers give birth to congenitally infected infants, 70% of whom are asymptomatic. *Remington and Desmonts* [71] have provided a comprehensive review of toxoplasmosis. It is unlikely that this infection is a significant cause of spontaneous abortion or recurrent reproductive failure [45, 90].

Toxoplasma cysts have been found in placental amnion and amniotic fluid [4, 91]. When congenital toxoplasmosis occurs in twins, the first born

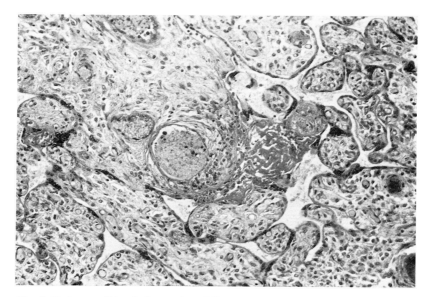

Fig. 4. Endovasculitis of placental syphilis. Perivascular villous inflammatory cell infiltrates are present at the central aspect of this illustration. The other villi show a severe increase of stromal cells. HE. × 200.

is more severely affected than the second and there are instances in which the second born is clinically normal [27]. Congenital toxoplasmosis may therefore be acquired by ascending intrauterine infection. Placental histopathologic changes, thus far reported, have drawn attention to a hematogenous route of infection [20, 26, 71]. These changes are similar to those of erythroblastosis but additionally include signs of infection. Generalized placental dysmaturity [9], numerous fetal erythroid cell precursors and focal villitis are typical and the cysts are pathognomonic. Additionally described funisitis and deciduitis are further evidence of ascending intrauterine infection [26].

Rubella

With regard to historic epidemics of rubella, there is a noteworthy numerical deficiency of placental histopathology reports. Severe placentitis has been described in a couple of placentas from full-term gestations [21]. Typical features of rubella placentitis include villous endothelial necrosis, perivasculitis, lymphohistiocytic villitis, karyorrhectic change of Hofbauer or stromal cells, calcific villous vascular changes and focal villous hemosiderin deposits [31, 69, 94]. These changes are manifestations of transplacental infection. *Seppala and Vaheri* [80], however, have drawn attention to the possibility that embryos may become infected with rubella via the genital tract; they isolated rubella virus from the cervix in 2 of 3 nonpregnant women, 2–6 days after the onset of clinical rubella infection.

Syphilis

Congenital syphilis is common in third-world countries [50]. In the United States, epidemiologic data are difficult to interpret, because of the small number of cases and diagnostic uncertainty of the disease [43]. Clinicopathologic observations and placental findings have been reported and reviewed [68, 73]. Syphilitic placentas tend to be large because of stromal cell proliferation and occasionally because of hydrops. Typical findings include diffuse dysmaturity, focal lymphoplasmacytic villitis, acute and chronic fetal endovasculitis (fig. 4), villous vascular obliteration and spirochetes. Warthin-Starry stains optimally show this but the organisms are very sensitive to antibiotics and may therefore not be demonstrable [73]. Spirochetes have been identified in placentas of 9 and 10 weeks' gestation [40].

Group B Streptococcus (Streptococcus agalactiae)

In American and British medical centers, the incidence of group B streptococcal sepsis has been estimated to be between 1.3 and 3 per 1,000 livebirths [12]. Because the overall fatality rate is 44% [12], prompt diagnosis and treatment are essential.

Amniotic and chorionic placental smears may be helpful; according to *Blanc* [26], the finding of numerous gram-positive streptococci or gram-positive rods should be interpreted as group B streptococcus or listeria, respectively, until proven otherwise. Amnionitis has been found in 15 of 20 placentas from newborns with group B β-hemolytic streptococcal infection [16]. In placentas I have examined, chorioamnionitis with streptococci occurred in less than 30%. Whereas respiratory and meningitic signs of sepsis have been manifest within the first few days of life, delayed meningitic manifestations have been reported in infants of several weeks age [13, 34].

Five serotypes have been identified (Ia, Ib, Ic, II and III). Because the prevalence of these serotypes may shift with time and place, specific associations between serotype and human virulence should be viewed cautiously [12]. Incomplete knowledge of epidemiology, placental histopathology, pathogenesis and immunology preclude conclusive recommendations of optimal prophylaxis [90].

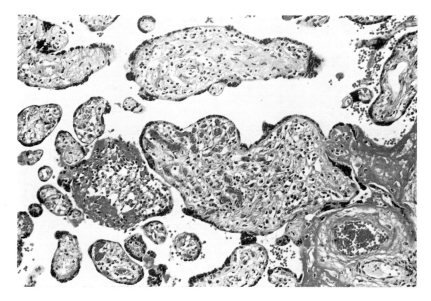

Fig. 5. Villitis of unknown etiology. Three villi in this picture show obvious inflammatory cell infiltrates. Of these, necrotizing change is present within the villus towards the lower left of the illustration. HE. ×200.

Villitis of Miscellaneous Intrauterine Infections

Pathologic aspects of many specific villitides have been described and reviewed [8, 18, 26]. There is a most important group of placental villitides, however, whose specific causes remain obscure [7, 33, 75–77]. From a morphologic standpoint, these entities are infectious (fig. 5). Biopsy and autopsy experience confirm this opinion. Viruses, chlamydia, mycoplasma and anaerobic bacteria may be causative. Until such time as pathologists, obstetricians, pediatricians and microbiologists elucidate these villitides, the scope of their pathologic significance will remain uncertain.

References

1 Alexander, E.F.: Maternal and neonatal infection with cytomegalovirus in Taiwan. Pediat. Res. *1:* 210 (1967).
2 Alford, C.A.; Reynolds, D.W.; Stagno, S.: Current concepts of chronic perinatal infections; in Modern perinatal medicine, pp. 285–306 (Year Book Medical Publishers, Chicago 1974).

3 Altshuler, G.: Placental villitis of unknown etiology: harbinger of serious disease? A four months' experience of nine cases. J. reprod. Med. *11:* 215–222 (1973).
4 Altshuler, G.: Toxoplasmosis as a cause of hydranencephaly. Am. J. Dis. Child. *125:* 251–252 (1973).
5 Altshuler, G.: Pathogenesis of congenital herpesvirus infection. Case report including a description of the placenta. Am. J. Dis. Child. *127:* 427–429 (1974).
6 Altshuler, G.: Immunologic competence of the immature human fetus. Morphologic evidence from intrauterine cytomegalovirus infection. Obstet. Gynec., N.Y. *43:* 811–816 (1974).
7 Altshuler, G.; Russell, P.: The human placental villitides. A review of chronic intrauterine infection. Curr. Top. Pathol. *60:* 63–112 (1975).
8 Altshuler, G.: Placentitis, with a new light on an old ToRCH; in Obstetrics and gynecology annual, vol. 6, pp. 197–221 (Appleton Century Crofts, Middlesex 1977).
9 Altshuler, G.: The placenta, how to examine it, its normal growth and development; in Perinatal diseases, pp. 5–22 (Williams & Wilkins, Baltimore 1981).
10 Amstey, M.S.: Current concepts of herpesvirus infection in the woman. Am. J. Obstet. Gynec. *117:* 717–725 (1973).
11 Amstey, M.S.: Neonatal risk following late gestational genital herpesvirus hominis infection. Am. J. Dis. Child. *129:* 985–990 (1975).
12 Anthony, B.F.; Okada, D.M.: The emergence of group B streptococci in infections of the newborn infant. A. Rev. Med. *28:* 355–369 (1977).
13 Baker, C.J.; Barrett, F.F.; Gordon, R.C.; Gow, M.D.: Suppurative meningitis due to streptococci of Lanefield group B: a study of 33 infants. J. Pediat. *82:* 724–729 1973).
14 Bale, J.F.; Reiley, T.T.; Bray, P.F.; Kelsey, D.K.: Cytomegalovirus and dual infections in infants. Archs Neurol., Chicago *37:* 236–238 (1980).
15 Ballard, R.A.; Drew, W.L.; Hufnagle, K.G.; Riedel, P.A.: Acquired cytomegalovirus infection in preterm infants. Am. J. Dis. Child. *133:* 482–485 (1979).
16 Becroft, D.M.O.; Farmer, K.; Mason, G.H.; Morris, M.C.: Perinatal infections by group B beta-haemolytic streptococci. Br. J. Obstet. Gynaec. *83:* 960–966 (1976).
17 Benirschke, K.: Routes and types of infection in the fetus and the newborn. Am. J. Dis. Child. *99:* 714 (1960).
18 Benirschke, K.; Driscoll, S.: The pathology of the human placenta (Springer, New York 1970).
19 Benirschke, K.; Driscoll, S,: The pathology of the human placenta, pp. 249 (Springer, New York 1970).
20 Benirschke, K,; Driscoll, S.: The pathology of the human placenta, pp, 274–278 (Springer, New York 1970).
21 Benirschke, K.; Driscoll, S.: The pathology of the human placenta, pp. 281–282 (Springer, New York 1970).
22 Benirschke, K.; Altshuler, G.: The future of perinatal physiopathology. Symposium on the functional physiopathology of the fetus and neonate, pp. 158–168 (Mosby, St. Louis 1971).
23 Benirschke, K.; Mendoza, G.R.; Bazeley, P.L.: Placental and fetal manifestations of cytomegalovirus infection. Virchows Arch. Abt. B Zellpath. *16:* 121–139 (1974).
24 Benirschke, K.: Personal communication (1981).

25 Blanc, W.A.: Pathways of fetal and early neonatal infection. J. Pediat. *59:* 473–495 (1961).
26 Blanc, W.A.: Pathology of the placenta, membranes, and umbilical cord in bacterial, fungal and viral infections in man; in Perinatal diseases, pp. 67–132 (Williams & Wilkins, Baltimore 1981).
27 Couvreur, J.; Desmonts, G.; Girre, J.Y.: Congenital toxoplasmosis in twins. J. Pediat. *89:* 235–240 (1976).
28 Creatsas, G.K.; Lolis, D.E.; Pavlatos, M.P.; Kaskarelis, D.B.: Bacteriology of amniotic fluid. Gynecol. obstet. Invest. *11:* 174–176 (1980).
29 Dehner, L.P.; Askin, F.B.: Cytomegalovirus endometritis. Report of a case associated with spontaneous abortion. Obstet. Gynec., N.Y. *45:* 211–214 (1975).
30 Desmonts, G.; Couvreur, J.: Congenital toxoplasmosis. New Engl. J. Med. *290:* 1110–1116 (1974).
31 Driscoll, S.G.: Histopathology of gestational rubella. Am. J. Dis. Child. *118:* 49–53 (1969).
32 Florman, A.L.; Gershon, A.A.; Blackett, P.R.; Nahmias, A.J.: Intrauterine infection with herpes simplex virus. Resultant congenital malformations. J. Am. med. Ass. *225:* 129–132 (1973).
33 Fox, H.: Pathology of the placenta (Saunders, London 1978).
34 Franciosi, R.A.; Knostman, J.D.; Zimmerman, R.A.: Group B streptococcal neonatal and infant infections. J. Pediat. *82:* 707–718 (1973).
35 Galask, R.P.; Snyder, I.S.: Antimicrobial factors in amniotic fluid. Am. J. Obstet. Gynec. *106:* 59–65 (1970).
36 Gibbs, R.S.; Castillo, M.S.; Rodgers, P.J.: Management of acute chorioamnionitis. Am. J. Obstet. Gynec. *136:* 709–713 (1980).
37 Goldman, R.L.: Herpetic inclusions in the endometrium. Obstet. Gynec., N.Y. *36:* 603–605 (1970).
38 Grossman, J.H.; Wallen, W.C.; Sever, J.L.: Management of genital herpes simplex virus infection during pregnancy. Obstet. Gynec., N.Y. *58:* 1–4 (1981).
39 Hanshaw, J.B.: Herpesvirus hominis infections in the fetus and newborn. Am. J. Dis. Child. *126:* 546–555 (1973).
40 Harter, C.A.; Benirschke, K.: Fetal syphilis in the first trimester. Am. J. Obstet. Gynec. *124:* 705–711 (1976).
41 Hayes, K.; Danks, D.M.; Gibas, H.: Cytomegalovirus in human milk. New Engl. J. Med. *287:* 177–178 (1972).
42 Herbst, A.L.: Coitus and the fetus. New Engl. J. Med. *301:* 1235–1236 (1979).
43 Kaufman, R.E.; Jones, O.G.; Blount, J.H.; Wiesner, P.J.: Questionnaire survey of reported early congenital syphilis: problems in diagnosis, prevention and treatment. Sex. Trans. Dis. *4:* 135–139 (1977).
44 Keenan, W.J.; Steichen, J.J.; Mahmood, K.; Altshuler, G.: Placental pathology compared with clinical outcome. A retrospective blind review. Am. J. Dis. Child. *131:* 1224–1228 (1977).
45 Kimball, A.C.; Kean, B.H.; Fuchs, F.: The role of toxoplasmosis in abortion. Am. J. Obstet. Gynec. *111:* 219–226 (1971).
46 Lang, D.J.; Kummer, J.F.; Hartley, D.P.: Cytomegalovirus in semen. Persistence and demonstration in extracellular fluids. New Engl. J. Med. *291:* 121–123 (1974).

47 Lapinleimu, K.; Koskimies, O.; Cantell, K.; Saxen, L.: Association between maternal herpesvirus infections and congenital malformations. Lancet *I:* 1127–1129 (1974).
48 Lee, F.K.; Nahmias, A.J.; Stagno, S.: Rapid diagnosis of cytomegalovirus infection in infants by electron microscopy. New Engl. J. Med. *299:* 1266–1270 (1978).
49 Lynn, E.J.: Letter to the editor. Teratology *19:* 267 (1979).
50 Marboe, C.; Tafari, N.; Judge, D.; Naeye, R.: Congenital syphilis: anthropometry and histologic analysis. Lab. Invest. *34:* 19 (1976).
51 Medearis, D.N., Jr.: Observations concerning human cytomegalovirus infection and disease. Bull. Johns Hopkins Hosp. *114:* 181 (1964).
52 Mills, J.L.; Harlap, S.; Harley, E.E.: Should coitus late in pregnancy be discouraged? Lancet *II:* 136–138 (1981).
53 Montgomery, J.R.; Flanders, R.W.; Yow, M.D.: Congenital anomalies and herpesvirus infection. Am. J. Dis. Child. *126:* 364–366 (1973).
54 Montgomery, R.; Youngblood, L.; Medearis, D.N.: Recovery of cytomegalovirus from the cervix in pregnancy. Pediatrics, Springfield *49:* 524–531 (1972).
55 Montplaisir, S.; Belloncik, S.; Leduc, N.P.; Onji, P.A.; Martineau, B.; Kurstak, E.: Electron microscopy in the rapid diagnosis of cytomegalovirus: ultrastructural observation and comparison of methods of diagnosis. J. infect. Dis. *125:* 533–538 (1972).
56 Naeye, R.L.; Dellinger, W.S.; Blanc, W.A.: Fetal and maternal features of antenatal bacterial infections. J. Pediat. *79:* 733–739 (1971).
57 Naeye, R.L.; Blanc, W.A.: Unfavorable outcome of pregnancy: repeated losses. Am. J. Obstet. Gynec. *116:* 1133–1137 (1973).
58 Naeye, R.L.; Peters, E.C.: Amniotic fluid infections with intact membranes leading to perinatal death: a prospective study. Pediatrics, Springfield *61:* 171–177 (1978).
59 Naeye, R.L.: Causes of perinatal mortality in the US Collaborative Perinatal Project. J. Am. med. Ass. *238:* 228–229 (1977).
60 Naeye, R.L.: Amniotic fluid infections, neonatal hyperbilirubinemia, and psychomotor impairment. Pediatrics, Springfield *62:* 497–503 (1978).
61 Naeye, R.L.: Coitus and associated amniotic-fluid infections. New Engl. J. Med. *301:* 1198–1200 (1979).
62 Naeye, R.L.; Peters, E.C.: Causes and consequences of premature rupture of fetal membranes. Lancet *I:* 192–194 (1980).
63 Nahmias, A.J.; Alford, C.A.; Korones, S.B.: Infection of the newborn with herpesvirus hominis. Adv. Pediat. *17:* 185–226 (1970).
64 Nahmias, A.J.; Roizman, B.: Infection with herpes-simplex viruses 1 and 2. New Engl. J. Med. *289:* 781–789 (1973).
65 Nahmias, A.J.: The ToRCH complex. Hosp. Pract. *May:* 65 (1974).
66 Naib, Z.M.; Nahmias, A.J.; Josey, W.E.; Wheeler, J.H.: Association of maternal genital herpetic infection with spontaneous abortion. Obstet. Gynec., N.Y. *35:* 260–263 (1970).
67 Numazaki, Y.; Yano, N.; Morizuka, T.: Primary infection with human cytomegalovirus; virus isolation from healthy infants and pregnant women. Am. J. Epidem. *91:* 410–417 (1970).
68 Oppenheimer, E.H.; Hardy, J.B.: Congenital syphilis in the newborn infant: clinical and pathological observations in recent cases. Johns Hopkins med. J. *129:* 63–82 (1971).

69 Ornoy, A.; Segal, S.; Nishmi, M.; Simcha, A.; Polishuk, W.Z.: Fetal and placental pathology in gestational rubella. Am. J. Obstet. Gynec. *116:* 949–956 (1973).
70 Panjvani, Z.F.K.; Hanshaw, J.B.: Cytomegalovirus in the perinatal period. Am. J. Dis. Child. *135:* 56–60 (1981).
71 Remington, J.S.; Desmonts, G.: Toxoplasmosis; in Infectious diseases of the fetus and newborn infant, pp. 191–332 (Saunders, Philadelphia 1976).
72 Reynolds, D.W.; Stagno, S.; Hosty, T.S.; Tiller, M.; Alford, C.A.: Maternal cytomegalovirus excretion and perinatal infection. New Engl. J. Med. *289:* 1–6 (1973).
73 Russell, P.; Altshuler, G.: The placental pathology of congenital syphilis: a neglected aid to diagnosis. Am. J. Dis. Child. *128:* 160–163 (1974).
74 Russell, P.: Inflammatory lesions of the human placenta. I. Clinical significance of acute chorioamnionitis. Am. J. Diag. Gynec. Obstet. *1:* 127–137 (1979).
75 Russell, P.: Inflammatory lesions of the human placenta. II. Villitis of unknown etiology in perspective. Am. J. Diag. Gynec. Obstet. *1:* 339–346 (1979).
76 Russell, P.: Inflammatory lesions of the human placenta. III. The histopathology of villitis of unknown aetiology. Placenta *1:* 227–244 (1980).
77 Sander, C.H.: Hemorrhagic endovasculitis and hemorrhagic villitis of the placenta. Archs Pathol. Lab. Med. *104:* 371–373 (1980).
78 Schacter, J.: Chlamydial infections. New Engl. J. Med. *298:* 490–495 (1978).
79 Schlievert, P.; Johnson, W.; Galask, R.P.: Bacterial growth inhibition by amniotic fluid. Am. J. Obstet. Gynec. *125:* 899–905 (1976).
80 Seppala, M.; Vaheri, A.: Natural rubella infection of the female genital tract. Lancet *ii:* 46–47 (1974).
81 Sever, J.L.: The prevention of mental retardation through control of infectious diseases. Publ. Hlth Serv. Publication, No. 1692, pp. 37–68 (Washington, 1966).
82 Siegel, J.D.; McCracken, G.H.: Sepsis neonatorum. New Engl. J. Med. *304:* 642–647 (1981).
83 South, M.A.; Tompkins, W.A.F.; Morris, C.R.; Rawls, W.E.: Congenital malformation of the central nervous system associated with genital type (type 2) herpesvirus. J. Pediat. *75:* 13–18 (1969).
84 Stagno, S.; Reynolds, D.W.; Huang, E.S.; Thames, S.D.; Smith, R.J.; Alford, C.A.: Congenital cytomegalovirus infection. Occurrence in an immune population. New Engl. J. Med. *296:* 1254–1258 (1977).
85 Stagno, S.: Congenital toxoplasmosis. Am. J. Dis. Child. *134:* 635–637 (1980).
86 Stagno, S.; Pass, R.F.; Reynolds, D.W.; Moore, M.A.; Nahmias, A.J.; Alford, C.A.: Comparative study of diagnostic procedures for congenital cytomegalovirus infection. Pediatrics, Springfield *65:* 251–257 (1980).
87 Stagno, S.; Reynolds, D.W.; Pass, R.F.; Alford, C.A.: Breast milk and the risk of cytomegalovirus infection. New Engl. J. Med. *302:* 1073–1076 (1980).
88 St. Geme, J.W.; Bailey, S.R.; Koopman, J.S.; Oh, W.; Hobel, C.J.; Imagawa, D.T.: Neonatal risk following late gestational genital herpesvirus hominis infection. Am. J. Dis. Child. *129:* 342–343 (1975).
89 Strawn, E.Y.; Scrimenti, R.J.: Intrauterine herpes simplex infection. Am. J. Obstet. Gynec. *115:* 581–582 (1973).
90 Stray-Pedersen, B.; Lorentzen-Styr, A.M.: Uterine toxoplasma infections and repeated abortions. Am. J. Obstet. Gynec. *128:* 716–721 (1977).

91 Stray-Pedersen, B.: Infants potentially at risk for congenital toxoplasmosis. Am. J. Dis. Child. *134:* 638–643 (1980).
92 Tafari, N.; Ross, S.M.; Naeye, R.L.; Galask, R.P.; Zaar, B.: Failure of bacterial growth inhibition by amniotic fluid. Am. J. Obstet. Gynec. *128:* 187 (1977).
93 Tejani, N.; Klein, S.W.; Kaplan, M.: Subclinical herpes simplex genitalis infections in the perinatal period. Am. J. Obstet. Gynec. *135:* 547 (1979).
94 Tondury, G.T.; Smith, D.W.: Fetal rubella pathology. J. Pediat. *68:* 867–879 (1966).
95 Torphy, D.E.; Ray, C.G.; McAlister, R.; Du, J.N.H.: Herpes simplex virus infection in infants: a spectrum of disease. J. Pediat. *76:* 405–408 (1970).
96 Wenckebach, G.F.C.; Curry, B.: Cytomegalovirus infection of the female genital tract. Archs Pathol. Lab. Med. *100:* 609–612 (1976).
97 Whitley, R.J.; Brasfield, D.; Reynolds, D.W.; Stagno, S.; Tiller, R.E.; Alford, C.A.: Protracted pneumonitis in young infants associated with perinatally acquired cytomegaloviral infection. J. Pediat. *89:* 16–22 (1976).
98 Whitley, R.J.; Nahmias, A.J.; Visintine, A.M.; Fleming, C.L.; Alford, C.A.: The natural history of herpes simplex virus infection of mother and newborn. Pediatrics, Springfield *66:* 489–494 (1980).
99 Wilson, C.B.; Remington, J.S.: What can be done to prevent congenital toxoplasmosis? Am. J. Obstet. Gynec. *138:* 357–363 (1980).
100 Witzleben, C.L.; Driscoll, S.G.: Possible transplacental transmission of herpes simplex infection. Pediatrics, Springfield *36:* 192–199 (1965).

Prof. Geoffrey Altshuler, Pathology Service, Oklahoma Children's Memorial Hospital, 940 NE 13th, Rm. 3B400, PO Box 26307, Oklahoma City, OK 73126 (USA)

Human Placental Lactogen and Ultrasonic Screening for the Detection of Placental Insufficiency

S. Kullander, K. Maršál, P.-H. Persson

Department of Obstetrics and Gynecology, General Hospital, University of Lund, Malmö, Sweden

Introduction

In obstetric practice, birth weight has always been considered an important factor for the prognosis of the neonate. According to *Gruenwald* [1], the pathogenesis of a small size at birth is, apart from preterm delivery, related to a reduced growth potential, or to deprivation resulting from an insufficient supply line. The concept of relating the growth of the fetus to gestational age is very important. Small-for-date fetuses incur the risk of increased perinatal mortality due to asphyxia but also suffer increased morbidity with irreversible neurological and intellectual sequelae [2–4]. The neonate of low birth weight due to preterm delivery suffers from organ immaturity during the change from intrauterine to extrauterine life. Once it has overcome the perinatal period, the prognosis is good [5]. The situation is more complicated for the growth-retarded fetus and the prognosis is less good [6]. To prevent persistent damage to developing organ systems, it is essential to diagnose intrauterine growth retardation (IUGR) at an early stage. Important maternal factors associated with IUGR and placental insufficiency include severe toxemia, hypertension, renal disease and intrauterine infections. However, often there are no clinical warning signs of IUGR or placental insufficiency. This emphasizes the need for suitable methods to screen and find pregnancies at risk for developing IUGR. Our understanding of the relationship between the fetus and the placenta is rather limited. There are very few parameters that directly measure placental function. Therefore, it is simpler to detect the result of placental insufficiency, namely the growth-retarded fetus. Though adequate, it is therefore not

Table I. Validity of a diagnostic test

Sensitivity (= tracing rate), %	probability to obtain a positive result in diseased subjects	$\dfrac{TP}{TP+FN} \times 100$
Specificity, %	probability to obtain a negative result in nondiseased subjects	$\dfrac{TN}{TN+FP} \times 100$
Predictive value of a positive test	probability of being diseased when the test is positive	$\dfrac{TP}{TP+FP} \times 100$
Predictive value of a negative test	probability of being nondiseased when the test is negative	$\dfrac{TN}{TN+FN} \times 100$
At-risk group, %		$\dfrac{TP+FP}{n} \times 100$

TP = True positives; FP = false positives; TN = true negatives; FN = false negatives; n = TP + FP + TN + FN = total population.

meaningful, to maintain a distinction between the concept of placental insufficiency and IUGR when evaluating clinical methods for the detection of placental insufficiency.

All screening methods must fulfil certain criteria: they must be simple, harmless, inexpensive and have a high acceptance. The size of the at-risk group detected at screening should not exceed the prevalence of the disease in the population too much [7] and must be such that it can be handled clinically. The diagnostic method should be evaluated in terms of sensitivity, specificity and predictive value of a negative or positive test [8, 9] (table I). The efficiency of a screening method can be improved by preselection. This can be done either by combining several screening methods or by repeated examinations.

This review evaluates the human placental lactogen (HPL) and ultrasound as screening methods for detection of IUGR, based on our own experience and on studies published in the literature. For simplicity we have calculated the tracing rate of each test, which corresponds to its sensitivity; furthermore, the size of the at-risk group selected by the test is given. The at-risk group comprises all pregnancies with positive tests, i.e. both true- and false-positive tests. The at-risk group should always be related to the prevalence of the disease in the whole population; therefore, this figure is also given.

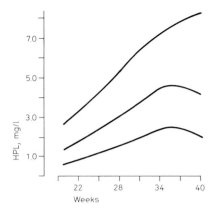

Fig. 1. HPL levels in maternal serum in normal pregnancies. Mean ± 2 SD.

HPL Screening

HPL is a polypeptide, chemically and biologically similar to human growth hormone [10, 11], produced by the syncytial trophoblast [12]. Maternal serum levels of HPL change with gestational age (fig. 1) and correlate grossly with placental weight and fetal weight [13]. In our population, however, HPL was poorly correlated with placental weight (r = 0.28), fetal weight (r = 0.26) and maternal serum estriol levels (r = 0.23). No correlations were found between the HPL serum levels and the time of the day or maternal meals [14]. The fact that HPL reflects the functioning trophoblastic mass makes it tempting to apply this hormone as a measure of placental function in a screening procedure. Table II gives the results of HPL screening for the detection of fetal distress and IUGR in six recent studies [15–20]. Only the studies by *Nielsen* et al. [15] and *Zlatnik* et al. [17] evaluated prospectively the screening of unselected normal pregnant populations. All six reports showed a very low tracing rate of a disorder. The studies [16, 18–20] with a preselection of at-risk fetuses (prevalence ≥ 10%) had generally higher tracing rates than the studies without preselection [15, 17]. In all the presented studies the detected at-risk group was of an acceptable size compared with the prevalence of the disorder. The tracing rate was, however, far too low to allow the HPL test to be recommended as a screening procedure. Furthermore, the presented studies had HPL determinations performed every week or every second week during the last trimester, which is too often and too expensive to be used in a general screening.

Table II. Results of HPL screening published in the literature

Disorder	Reference	Material		Results	
		n	prevalence of disorder %	tracing rate of disorder %	at-risk group %
Birth weight < 10th centile	15	1,660	6	13	2
Birth weight < 10th centile	16	148	17	41	15
IUGR	17	806	2	7	6
Birth weight < 5th centile + stillbirths and neon. deaths	18	972	10	32	12
High-risk pregnancy	19	1,029	10	15	6
Fetal distress	20	547	22	26	12

High-risk pregnancy defined as: birth weight < 5th centile, perinatal deaths, 1-min Apgar score ≤ 3, congenital abnormalities.
Fetal distress defined as: fetal heart rate variations, meconium-stained liquor, 1-min Apgar score ≤ 6, birth weight ≤ 2,500 g.

Even though the HPL test does not seem to have a place in an antepartum general screening for IUGR, many studies have confirmed that HPL can predict fetal well-being quite accurately when an adequate preselection is performed. Such a preselection criterion can be a bad obstetric history and/or a clinical complication of the present pregnancy, e.g. hypertension or IUGR. *Spellacy* et al. [21] showed convincingly that by recognizing pathological HPL values in a preselected high-risk population, the perinatal loss could be reduced from 15 to 3.4%. Several other studies have confirmed that HPL could be used for screening out the risk patients in pregnancies preselected because of hypertension [16, 20–22]. No reports have been published on the predictive value of HPL screening after preselection because of IUGR detected by ultrasound.

The theoretical basis for using HPL measurements as a test of placental function is well founded. However, until now, no study could convincingly demonstrate a clinical value of HPL measurements for the detection of placental insufficiency. This might be due to the lack of a suitable independent method to measure the placental function for comparison with the HPL test. IUGR and fetal asphyxia have been used as clinically relevant criteria; this probably gave rise to misleading results, as many cases of IUGR or asphyxia are not related to placental insufficiency.

Ultrasound Screening

Ultrasound fetal biometry started with the finding by *Willocks and Densmore* [23] that dysmature infants had on average lower ultrasonically measured biparietal diameters (BPD) than normal infants. They also showed a correlation between BPD and fetal weight. However, in the individual fetus, the weight-predictive capacity of BPD measurement was not better than clinical palpation [24]. By including measurements of other dimensions than BPD in fetometry, e.g. trunk measurements, *Thompson* et al. [25] found that fetal weight could be estimated more accurately.

The ultrasonic identification of a small fetus is not directly indicative of fetal growth retardation. In the Malmö population, 80% of the third-trimester fetuses with BPD below the 5th centile of the reference values were explained by miscalculated dates of confinement. Possible IUGR cannot be differentiated from an erroneous estimate of gestational age by serial ultrasound measurement starting in the last trimester [26]. Therefore, a reliable estimate of gestational age is of paramount importance in the diagnosis of IUGR by ultrasound. Several studies have shown that approximately 25% of the pregnant women cannot give a reliable date of their last menstrual period [27, 28]. In fact, in as many as 15% of the pregnancies, the gestational age is miscalculated by more than 14 days [29]. To allow detection of IUGR in pregnancy, the ultrasonic estimation of gestational age must be performed well before the time when the growth retardation becomes manifest. This implies a general screening of the pregnant population in the first half of pregnancy.

The ultrasonic fetometry parameters are given in table III. Of those, crown-rump-length (CRL), BPD and the length of the fetal femur are suitable for the estimation of gestational age. In 90% of the cases, the CRL-estimated gestational age agrees with the true gestational age within 3 days and the BPD-estimated gestational age within 5 days [30, 31]. According to *O'Brien* et al. [32], femur measurements are comparable with BPD measurements for the estimation of gestational age.

To monitor fetal growth, several ultrasound methods have been suggested in the literature. Not all of these methods are suitable for use in a general screening. The ultrasound equipment must be simple to handle and must have a sufficient technical quality. Modern real-time scanners, preferably the linear array scanners, fulfil satisfactorily these demands. Therefore, the fetometric parameters to be used in a screening should be measurable by the real-time technique. The formulas for fetometry should be

Table III. Ultrasonic biometric parameters used for the assessment of fetal well-being

Fetal head	area
	circumference
	diameter(s) (BPD, fronto-occipital diameter)
Fetal trunk	diameter(s)
	circumference
	area
	volume
Fetal crown-rump-length	
Fetal femur length	
Combinations and ratios of above parameters	
Total intrauterine volume	
Placenta	volume
	area

simple not necessitating computers as these are not generally available. For the prediction of fetal weight, most authors combine several parameters in formulas; most developed formulas assess fetal weight with a standard error (SD) of 7–10% of the true weight [33–36]. Only few formulas have been applied in a general screening for the detection of IUGR [37, 38].

A routine screening program for the entire pregnant population was introduced at the Department of Obstetrics and Gynecology in Malmö in October 1973. The primary purpose of the screening was to improve the detection rate of twin pregnancies. The screening program began with one examination in the 30th week. Later, the program was altered in several steps; at present, all pregnant women are offered ultrasonic examination in the 17th and 33rd postmenstrual week. 97% of the pregnant women participate in the screening. At the first examination, an estimate of gestational age is based solely on the BPD measurement and the gestational age is adjusted to the mean gestational day for the BPD value obtained. The second examination is performed in the 33rd week and aims at detecting deviations of fetal growth (positive or negative), fetal malformations, and describing the placental implantation more precisely. Up to 1978, the assessment of fetal growth entirely relied on serial BPD measurements [26]. In 1979, a new mathematical model including measurements of other fetal dimensions was developed. The prerequisite for this model was the use of real-time linear array ultrasound equipment. The new formula predicted weight deviation by combined measurements of the BPD and the mean of two perpendicular abdominal diameters (AD) (fig. 2) in the 33rd week.

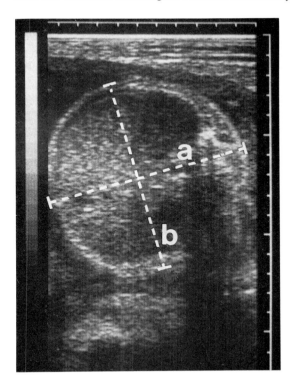

Fig. 2. Real-time image of a transverse section through fetal abdomen at the level of the umbilical vein. AD is the mean of the anteroposterior (a) and the transverse diameter (b).

$$AD = \frac{a+b}{2}.$$

The measured values of these parameters were related to the mean value for the gestational age on the day of measurement. The difference between the found and the expected value characterized the fetal growth deviation. The following formula was developed:

D-weight = D-BPD × 1.109 + D-AD × 1.887 + 0.259.

D-BPD and D-AD correspond to the difference in millimeters between measured and expected value for the gestational age on the day of measurement. The D-weight is the deviation in percent from the expected, i.e. normal mean weight. The D-weight $\leq 22\%$ is equal to a growth below mean -2 SD and suspected of IUGR. We have used deviations from expected values for the gestational age rather than absolute values of weight

Table IV. Efficacy of ultrasound fetometry for the detection of birth weights below the normal mean – 2 SD in 524 consecutive pregnancies examined at 33 and 38 weeks

Measured parameter	Prevalence of IUGR, %	Tracing rate, %		At-risk group, %	
		33 weeks	38 weeks	33 weeks	38 weeks
BPD	3.1	39	59	9	13
AD	3.1	40	72	4	4
AD/BPD	3.1	11	20	3	2
D-weight (BPD + AD)	3.1	46	80	4	4

to describe fetal growth. The observer is then not restricted to examine the fetus ultrasonically at a certain fixed gestational age. If the examination is done 1 week earlier or later, the D-BPD and D-AD values will most likely remain unchanged.

Before accepting the formula for screening for IUGR, a pilot study was performed on 524 unselected pregnant women examined in the 17th, 33rd and 38th week. Four fetometric functions based on BPD and AD measurements were tested: BPD measurements alone, AD measurements alone, ratio AD/BPD and the newly developed formula giving the D-weight. Values below mean – 2 SD were considered indicative of IUGR. The results of this pilot study are given in table IV.

Based on the pilot study, the D-weight concept was applied for general screening. By the formula, the birth weight can be predicted with an error (SD) of 9% in the 33rd week, 8% in the 38th week, and 7% immediately before delivery. 92% of the IUGR were detected if the last examination was performed within 3 weeks before delivery. At longer intervals from the measurement to delivery, the tracing rate decreased. Fetometry every 3rd week in the last trimester detects nearly all cases of growth retardation. To detect the majority of growth retardations (approx. 80%), it is necessary to examine all pregnant women twice in the last trimester of pregnancy, e.g. in the 33rd and in the 37th gestational week. At present, a compromise is necessary in Malmö: all pregnancies are examined in the 33rd week and all fetuses with a predicted weight below mean – 0.5 SD are re-examined in the 35th and the 38th gestational week (fig. 3). The efficiency of this procedure to detect growth deviations has been evaluated for all pregnancies during 1981 in Malmö (table V). Our screening procedure detects 60% of the growth-retarded infants with few false-positive diagnoses.

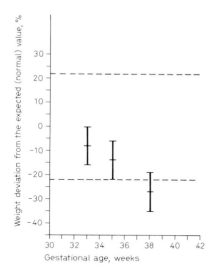

Fig. 3. Case of IUGR detected by ultrasound screening by calculating the weight deviation from the measurements of BPD and AD (see text). Ordinate: weight deviation from the expected (normal) value in percent; abscissa: gestational age in weeks. The weight deviations and the 75% confidence limits (vertical bars) are presented according to the measurements in the 33rd, 35th and 38th week. The newborn was delivered in the middle of the 40th week with a birth weight of 2,400 g, i.e. 30 percent below the normal mean weight for the corresponding day of gestation and below the 1st percentile of the Swedish standard population. The measurement in the 33rd week was inconclusive for IUGR; increasing predictive value of the additional measurements is demonstrated.

Discussion

In contrast to HPL, the screening by ultrasound has several spin-off effects with positive bearing on the clinical management of pregnancies (table VI). The ultrasound method does not indicate directly placental insufficiency, which is the cause of IUGR in only 10% [39]. There are two basic patterns of IUGR detected by ultrasound: symmetric IUGR with uniform reduction of fetal size and asymmetric IUGR where the fetal trunk is wasted but the head is spared. The latter pattern of IUGR is supposed to be caused by placental insufficiency [40, 41] and described as the common type. However, in our population, only 10–20% IUGR are asymmetric. For the detection of asymmetric IUGR, calculation of the ratio between the fetal head and trunk measurements was suggested [42]. *Varma* et al. [43]

Table V. Results of ultrasound screening for IUGR of entire pregnant population in Malmö in 1981. Prevalence of IUGR = 4.0%

	n		%
True positives	54	sensitivity (= tracing rate)	60
False positives	29	specificity	99
True negatives	2,109	predictive value of a positive test	65
False negatives	36	predictive value	98
Total	2,228	at-risk group	3.7

Table VI. Diagnoses and findings at ultrasound screening at 17 and 33 gestational weeks

	%
Unexpected miscalculated dates (>14 days)	15
IUGR	2.4
Multiple gestations	1
Intrauterine fetal demise	1
Fetal malformations	0.4
Low-implanted placenta	0.4
Placenta praevia	0.1
Hydatidiform mole	0.01
Miscellaneous	2

found the head to abdomen area ratio to be superior to other measurements in a preselected group of pregnancies clinically suspected of IUGR. To our knowledge, no prospective study has been published applying the ratio method to an unselected population. In our study (table V), the ratio between BPD and AD was not specific enough to be used as a screening method.

The literature contains many reports concerning antepartum detection of IUGR. Little attention has been paid to the assessment of placental size in utero. *Hoogland* et al. [44] measured prospectively the placental area. A small placental area in midgestation appeared to be of prognostic value in identifying at-risk groups for IUGR. Equipment for routine assessment of placental size is not yet available, however.

Ultrasound fetometry for screening for IUGR by ultrasonic fetometry searches for a late result of a pathological process, e.g. placental insufficiency. It would be desirable to diagnose the impaired function of the placenta before it seriously affected the fetus. This consideration stimulated several researchers to investigate some of the intrauterine dynamic functions of the fetus and to evaluate if changes in those functions would be useful for the identification of fetal health deterioration and of the changes in the function of the fetoplacental unit. The occurrence of fetal breathing movements and general fetal movements has been proved to be a normal phenomenon in pregnancy. Fetal motor functions are easily detectable by the real-time ultrasound technique. The reduction, cessation or change in the fetal activity pattern might indicate a loss of fetal normality. The idea of using fetal breathing movement examination as a clinical test was supported by the report on abolishment of breathing movements by hypoxia in fetal lambs [45]. However, when ultrasound examination of fetal breathing movements and fetal general movements was applied to the human fetus in a clinical situation, the predictive value of a positive test, i.e. the absence of normal motor activity, was very low [46, 47]. This is due to the very large variability in the incidence of fetal breathing movements and general movements and to the existence of physiological short-term and long-term cyclicity. Besides, the ultrasound real-time examinations of intrauterine fetal activity are very time-consuming and thus not applicable for the screening of a whole pregnant population.

Recently, a method was developed which makes it possible to directly measure blood flow in the umbilical vein, fetal descending aorta and fetal vena cava inferior [48]. The ultrasonic method for fetal blood flow measurement has been systematically evaluated in comparison with the electromagnetic method for blood flow measurements and found reliable [49]. The first reports on changes in the umbilical blood flow in pregnancies with IUGR have been published [50]. The method can hardly be used for the screening of all pregnancies but it seems that it might give valuable information on fetal blood circulation in cases of suspected placental insufficiency, detected at screening by other methods, e.g. ultrasound fetometry.

The ultrasound technique has many potentials that are not yet fully explored. One of the possible future applications of ultrasound is tissue characterization by analysis of the reflected ultrasound [51]. Preliminary results suggest that the liver and brain tissue of fetuses with serious IUGR has physical qualities different from those in normal-grown fetuses [*Persson*, unpubl.]. In the future, direct noninvasive evaluation of fetal and placental tissue characteristics will very probably be possible.

Ultrasound fetometry detects fetuses of small size and slow growth rate. Other methods must be applied to this preselected group of small fetuses to detect those who are at-risk for developing sequelae. Of biochemical tests it is likely that both HPL and estriol measurements can be valuable in the monitoring of pregnancies complicated with IUGR. Raised maternal α-fetoprotein (AFP) levels may result from a placental complication causing leakage of fetal AFP into the maternal circulation. We found that maternal serum AFP assessment has the ability to trace IUGR [52]. AFP screening can replace ultrasound if performed around 25 gestational weeks. After 30 weeks, the normal variation of AFP becomes too wide to allow the detection of complicated pregnancies. For the same reason, AFP cannot be used to monitor complicated pregnancies in the last trimester.

Ultrasound fetometry should also be compared with the simple measurement of the symphysis-fundal height (SF) according to *Westin* [53]. This method has a tracing rate of 62% and the at-risk group is 8% of the pregnancies [*Westin*, personal communication]. Both ultrasound fetometry and symphysis-fundal height measurements need an early ultrasound assessment of gestational age. The clinical value of the two methods is then apparently equal. However, with ultrasound, the results are obtained with very few (1.2) examinations in the last trimester, whereas symphysis-fundal height measurements have to be performed every week or every second week. Ultrasound is more expensive and requires much more training and education than symphysis-fundal height measurements. However, ultrasonic fetometry has a longer predictive interval and a more favorable relation between the predicted at-risk group and the prevalence of disease.

Conclusion

Placental insufficiency is most often detected first when it causes an impairment of fetal well-being, manifesting itself as IUGR or fetal asphyxia. In screening the total pregnant population, ultrasonic fetometry has proved to be the best method for identifying the fetuses at risk of developing IUGR. A reliable gestational age is a prerequisite for the evaluation of the fetal growth by ultrasound. Therefore, ultrasound screening necessitates an examination in the first half of pregnancy; the gestational age is then estimated within 5 days of the true age. Fetal growth can then be monitored at a second ultrasound examination in the last trimester of pregnancy. A newly developed formula using ultrasonic determination of simple linear measurements of the fetal head and abdomen enables the prediction of a deviation in the fetal growth later on in the pregnancy. In the 33rd week of pregnancy, 46% of the IUGR fetuses were predicted. If an additional ultrasound examination was performed, e.g. in the 38th week of pregnancy, the tracing rate of IUGR increased to 80%. Measurement of the HPL levels in the maternal serum has a low tracing rate of IUGR when used on an unselected pregnant population. However, HPL measurements give valuable information on fetal well-being when done on a preselected groups, e.g. on the at-risk pregnancies detected at ultrasound screening. The ultrasonic screening of an unselected obstetric population, including one measurement in early pregnancy and one or more measurements in late pregnancy, is an inexpensive and effective procedure, with several other positive effects, e.g. detection of fetal malformations and placenta insertion. The at-risk group selected by ultrasound fetometry must always be further evaluated and monitored. For this purpose, a combination of biochemical (HPL, estriol) and biophysical (CTG, subsequent ultrasound fetometry, ultrasonic measurements of fetal dynamic functions) methods seems at present preferable.

References

1 Gruenwald, P.: Pathology of the deprived fetus and its supply; in Elliott, Knight, Size at birth, p. 3 (Elsevier/Excerpta Medica/North-Holland, Amsterdam 1974).
2 Fitzhardinge, P.M.; Steven, E.M.: The small-for-date infant. II. Neurological and intellectual sequelae. Pediatrics, Springfield *50:* 50 (1972).
3 Sabel, K.G.; Olegård, R.; Victorin, L.: Remaining sequelae with modern perinatal care. Pediatrics, Springfield *57:* 652 (1976).
4 Finnström, O.; Nordgren, A.-M.; Svensson, A.; Oldfelt, V.: Låg födelsevikt och senare skolresultat. Läkartidningen *78:* 3684 (1981).
5 Rush, R.W.; Keirse, M.J.N.C.; Horvat, P.; Baum, J.D.; Andersson, A.B.M.; Turnbull, A.C.: Contribution of preterm delivery to perinatal mortality. Br. med. J. *ii:* 965 (1976).
6 Davie, T.; Butler, N.: From birth to seven, p. 21 (Longman, London 1972).
7 Nilsson, B.A.: Diagnos och behandling vid tvillinggraviditet. Läkartidningen *76:* 1045 (1979).
8 Vecchio, T.J.: Predictive value of a single diagnostic test in unselected populations. New Engl. J. Med. *274:* 1171 (1966).
9 Wulff, H.R.: Rational diagnosis and treatment (Blackwell, Oxford 1976).
10 Friesen, H.: Purification of a placental factor with immunological and chemical similarity to human growth hormone. Endocrinology *76:* 369 (1968).
11 Kaplan, S.L.; Grumbach, M.M.: Studies of a human and simian placental hormone with growth hormone-like and prolactin-like activities. J. Clin. Endocrinol. *24:* 80 (1964).
12 Beck, J.S.; Gordon, R.L.; Donald, D.: Characterization of antisera to a growth hormone-like placental antigen (human placental antigen). Immunofluorescence

studies with these sera on normal and pathological syncytiotrophoblast. J. Path. 97: 545 (1969).
13 Spellacy, W.N.; Carlson, K.L.; Birk, S.A.: Human placental lactogen levels as a variable of placental weight and infant weight. Am. J. Obstet. Gynec. 95: 118 (1966).
14 Vigneri, S.; Squatrito, S.; Pezzino, V.: Spontaneous fluctuations of human placental lactogen during normal pregnancy. J. clin. Endocr. Metab. 40: 506 (1974).
15 Nielsen, P.V.; Egebo, K.; Find, F.; Olsen, C.E.: Prognostic value of human placental lactogen (HPL) in an unselected obstetrical population. Acta obstet. gynec. scand. 60: 469 (1981).
16 Morrison, M.B.; Green, P.; Oomen, B.: The role of human placental lactogen in antepartum fetal assessment. Am. J. Obstet. Gynec. 136: 1055 (1980).
17 Zlatnik, M.D.; Varner, M.W.; Hauser, K.S.: Human placental lactogen: a predictor of perinatal outcome? Obstet. Gynec., N.Y. 54: 205 (1979).
18 Letchworth, A.T.; Slattery, M.; Dennis, K.J.: Clinical application of human-placental-lactogen values in late pregnancy. Lancet i: 955 (1978).
19 Gordon, Y.B.; Lewis, J.P.; Pendlebury, D.J.; Leighton, M.; Gold, J.: Is measurement of placental function and maternal weight worthwhile? Lancet i: 1001 (1978).
20 England, P.; Lorrimer, D.; Fergusson, J.C.; Moffat, A.; Kelly, A.: Human placental lactogen: the watchdog of fetal distress. Lancet i: 5 (1974).
21 Spellacy, M.D.; Buhi, W.C.; Birk, S.A.: The effectiveness of human placental lactogen measurements as an adjunct in decreasing perinatal deaths. Am. J. Obstet. Gynec. 121: 835 (1975).
22 Ylikorkala, O.: Maternal serum HPL levels in normal and complicated pregnancy as an index of placental function. Acta obstet. gynec. scand., suppl. 26, p. 40 (1973).
23 Willocks, J.; Densmore, I.R.: Assessment of gestational age and prediction of dysmaturity by ultrasonic fetal cephalometry. J. Obstet. Gynaec. Br. Commonw. 78: 804 (1971).
24 Beazley, J.M.; Kurjak, A.: Prediction of fetal maturity and birth weight by abdominal palpation. Nursing Times June: 14 (1973).
25 Thompson, H.E.; Holmes, J.H.; Gottesfeld, K.R.; Taylor, E.S.: Fetal development as determined by pulse echo technique. Am. J. Obstet. Gynec. 92: 44 (1965).
26 Persson, P.-H.; Grennert, L.; Gennser, G.: Diagnosis of intrauterine growth retardation by serial ultrasonic cephalometry. Acta obstet. gynec. scand., suppl. 78, p. 40 (1978).
27 Grennert, L.; Persson, P.-H.; Gennser, G.: Benefits of ultrasonic screening of a pregnant population. Acta obstet. gynec. scand., suppl. 78, p. 5 (1978).
28 Campbell, S.: The antenatal assessment of fetal growth and development: the contribution of ultrasonic measurement; in Roberts, Thompson, The biology of human fetal growth, p. 15 (Taylor & Francis, London 1976).
29 Hansmann, M.: 4th Eur. Congr. on Ultrasound in Medicine, Dubrovnik 1981.
30 Robinson, H.P.: Sonar measurements of fetal crown-rump-length as means of assessing maturity in first trimester of pregnancy. Br. med. J. iv: 28 (1973).
31 Persson, P.-H.; Grennert, L.; Gennser, G.: Normal range curves for the intrauterine growth of the biparietal diameter. Acta obstet. gynec. scand., suppl. 78, p. 15 (1978).
32 O'Brien, G.D.; Queenan, J.T.; Campbell, S.: Assessment of gestational age in the

second trimester by real-time ultrasound measurement of the femur length. Am. J. Obstet. Gynec. *139:* 540 (1981).
33 Lunt, R.; Chard, T.: A new method for estimation of fetal weight in late pregnancy by ultrasonic scanning. Br. J. Obstet. Gynaec. *83:* 1 (1976).
34 Warsof, S.L.; Gohari, P.; Berkowitz, R.L.; Hobbins, J.C.: The estimation of fetal weight by computer-assisted analysis. Am. J. Obstet. Gynec. *128:* 881 (1977).
35 McCallum, W.D.; Brinkley, J.F.: Estimation of fetal weight from ultrasonic measurements. Am. J. Obstet. Gynec. *133:* 195 (1979).
36 Eik-Nes, S.H.; Gröttum, P.; Andersson, N.J.: Estimation of fetal weight by ultrasound measurement. II. Clinical application of a new formula. Acta obstet. gynec. scand. (in press).
37 Wittman, B.K.; Robinson, H.P.; Aitchison, T.; Fleming, J.E.E.: The value of diagnostic ultrasound as a screening test for intrauterine growth retardation: comparison of nine parameters. Am. J. Obstet. Gynec. *134:* 30 (1979).
38 Neilson, J.P.; Whitfield, C.R.; Aitchison, T.C.: Screening for the small-for-dates fetus: a two-stage ultrasonic examination schedule. Br. med. J. *i:* 1204 (1980).
39 Gruenwald, P.: Chronic fetal distress and placental insufficiency. Biol. Neonate *5:* 215 (1963).
40 Crane, J.P.; Kopta, M.M.; Welt, S.I.; Sauvage, J.P.: Abnormal fetal growth patterns. Ultrasonic diagnosis and management. Obstet. Gynec., N.Y. *50:* 205 (1977).
41 Campbell, S.: Fetal growth. Clin. Obstet. Gynec. *1:* 41 (1974).
42 Campbell, S.; Thoms, A.: Ultrasound measurement of the fetal head to abdomen circumference ratio in the assessment of growth retardation. Br. J. Obstet. Gynaec. *84:* 165 (1977).
43 Varma, R.T.; Taylor, H.; Briges, C.: Ultrasound assessment of fetal growth. Br. J. Obstet. Gynaec. *86:* 623 (1979).
44 Hoogland, H.J.; de Haan, J.; Martin, C.B.: Placental size during early pregnancy and fetal outcome: a preliminary report of a sequential ultrasonographic study. Am. J. Obstet. Gynec. *138:* 441 (1980).
45 Dawes, G.S.: Breathing and rapid-eye-movement sleep before birth; in Comline, Cross, Dawes, Nathanielsz, Foetal and neonatal physiology. Proceedings of the Sir J. Barcroft Centenary Symposium, p. 49 (Cambridge University Press, Cambridge 1973).
46 Maršál, K.: Fetal breathing movements – characteristics and clinical significance. Obstet. Gynec., N.Y. *52:* 394 (1978).
47 Maršál, K.: Fetal movements and fetal breathing movements in the second half of pregnancy. A review of the clinical ultrasonic studies; in Kurjak, Kratochwil, Recent advances in ultrasound diagnosis 3. Excerpta Med. Int. Congr. Ser., No. 553, p. 174 (1981).
48 Eik-Nes, S.H.; Brubakk, A.D.; Ulstein, M.: Measurement of human fetal blood flow. Br. med. J. *i:* 283 (1980).
49 Eik-Nes, S.H.; Maršál, K.; Kristoffersen, K.; Vernersson, E.: Noninvasive Messung des fetalen Blutstromes mittels Ultraschall. Ultraschall Med. *2:* 226 (1981).
50 Gill, R.W.; Warren, P.S.; Griffiths, K.A.; Garrett, W.J.; Kossoff, G.: Umbilical blood flow in high risk pregnancy; in Kurjak, Kratochwil, Recent advances in ultrasound diagnosis 3. Excerpta Med. Int. Congr. Ser., No. 553, p. 220 (1981).
51 Lindström, K.; Holmer, N.-G.: Two methods for quantified ultrasonic tissue charac-

terization; in Kurjak, Recent advances in ultrasound diagnosis 2. Excerpta Med. Int. Congr. Ser., No. 498, p. 484 (1980).
52 Persson, P.-H.; Grennert, L.; Gennser, G.; Eneroth, P.: Fetal BPD growth and plasma concentrations of HPL, HCG, estriol, and AFP in normal and pathological pregnancy. Br. J. Obstet. Gynaec. *87:* 25 (1980).
53 Westin, B.: Schwangerschaftsüberwachung mittels Gravidogramm. Zentbl. Gynäk. *102:* 257 (1980).

Prof. S. Kullander, MD, Department of Obstetrics and Gynecology, General Hospital, University of Lund, S-214 01 Malmö (Sweden)

Placental Sulfatase Deficiency

Biochemical and Clinical Aspects

Tetsuya Nakayama, Takumi Yanaihara

Department of Obstetrics and Gynecology, School of Medicine,
Showa University, Tokyo, Japan

Introduction

Increasing amounts of estrogens are produced in the fetoplacental unit during pregnancy. As the fetus and the placenta are both involved in the biosynthesis of estrogen, determination of maternal urinary estriol excretion is a widely used method for assessing fetal well-being in high-risk pregnancies. During pregnancy, estriol, the major estrogen, is formed by a unique biosynthetic process which involves the fetal adrenal glands as a source of dehydroepiandrosterone sulfate (DHA-S), the fetal liver as a site of 16α-hydroxylation of DHA-S, and the placenta in which 16αOH-DHA-S is hydrolyzed and aromatized to form estriol (fig. 1).

Very low estriol excretion during the third trimester has been noted in the following clinical situations: (1) fetal death in utero; (2) anencephaly; (3) hypoplasia of fetal adrenal glands; (4) administration of excess doses of glucocorticoid to the mother, and (5) placental sulfatase deficiency.

The first four situations are considered to reflect an impairment of fetal adrenal function. However, a very small number of pregnancies have been reported in which the fetus was healthy in spite of the extremely low estriol excretion rates due to placental sulfatase enzyme deficiency.

Low estriol levels resulting from placental sulfatase deficiency were first reported by *France and Liggins* [1] in 1969. Since then, approximately 30 cases have been reported [2–18].

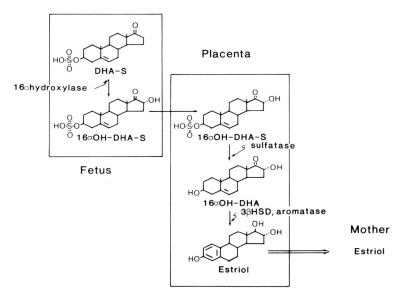

Fig. 1. Formation of estriol in the fetoplacental unit.

Biochemical Observations

Placental Enzyme Activities

Placental sulfatase deficiency is confirmed by postnatal in vitro incubation and/or perfusion studies. An essential step in the placental biosynthesis of estriol is the hydrolysis of 16αOH-DHA-S. 3β-Hydroxysteroid dehydrogenase (3β-HSD), an aromatizing enzyme system and 17β-hydroxysteroid dehydrogenase (17β-HSD) are required for further conversion of the free Δ^5-3βol steroid to form estriol.

Figure 2 shows the results of incubation studies in 4 cases of placental sulfatase deficiency reported by *Yanaihara* et al. [18]. Incubations were carried out in a total volume of 2 ml Tris buffer using 0.05 µCi labeled precursors and unlabeled steroids to give a final concentration of 10 nmol/flask. Sulfatase activity was expressed as the rate of hydrolysis of DHA-S during incubation with the 800 g supernatant of the placental homogenate. 3β-HSD activity was expressed as the rate of formation of progesterone from pregnenolone using the same tissue preparation as was used for sulfatase activity. To determine the aromatizing enzyme activity, androstenedione was incubated with placental homogenate, and the formation of estrogens measured.

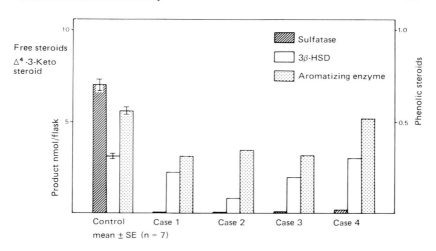

Fig. 2. Placental enzyme activities in 4 placentas suspected sulfatase deficiency. The results are expressed as the mean of duplicate measurements.

0.4 μmol of NAD and NADPH were added for the measurement of sulfatase activity and aromatase activity, while only NAD was used for 3β-HSD. The incubation was continued for 30 min at 37 °C in $O_2:CO_2$ (95:5) except for the measurement of 3β-HSD where CO gas was used. Enzyme activities were expressed as nanomoles/flask of the product. Compared with the controls, the placentas from all the 4 cases showed extremely low sulfatase activities. Except for case 2, there were no significant changes in 3β-HSD and aromatizing enzyme activities. Characteristic absence of sulfatase activity was demonstrated in all the cases in whom the enzyme activity for pregnenolone-sulfate and estrone-sulfate were also reported to be negligible [5, 6, 14]. Both the placental enzyme responsible for hydrolyzing steroid alcohol-3-sulfate and the enzyme activity for the hydrolysis of arylsulfates were found to be extremely low (4, 5, 7]. It was reported by *Shapiro* et al. [12] that the placental activity of arylsulfatase C was severely deficient, while the levels of arylsulfatase A and B were within the normal range for this tissue.

Oakey et al. [5] reported a case in whom placental sulfatase deficiency was associated with the reduction of all other enzymes concerned with the metabolism of free steroids (3β-HSD, aromatase, 17β-HSD), while others reported that these enzymes were not affected.

Table I. Steroid levels in maternal venous blood (mean ± SE; ng/ml)

Steroids	Placental sulfatase deficiency	n	Normal	n
Free steroid				
Pregnenolone	93.8	1	30.9 ± 4.8	10
Progesterone	175.3 ± 26.1	3	131.7 ± 19.5	10
16αOH-preg.	3.4 ± 0.5	2	2.7 ± 0.3	10
16αOH-prog.	2.8 ± 1.0	2	13.4 ± 2.7	10
20αOH-prog.	98.0	1	42.8 ± 6.3	10
DHA	8.1 ± 2.8	3	3.4 ± 0.7	10
16αOH-DHA	9.5 ± 1.8	3	10.7 ± 1.3	10
16αOH-A-dione	1.8	1	5.1 ± 0.5	5
16αOH-test.	0.3	1	8.0 ± 0.9	5
Estrone	1.4 ± 0.6	3	6.7 ± 1.7	10
Estradiol	5.8 ± 1.9	3	23.0 ± 2.2	9
Estriol	1.0 ± 0.4	3	11.6 ± 1.8	10
Estetrol	0.4 ± 0.1	2	1.4 ± 0.3	10
Conjugated steroid				
Pregnenolone	301.4 ± 68.6	2	152.6 ± 14.6	10
16αOH-preg.	266.3 ± 159.3	2	15.6 ± 3.4	10
DHA	405.2 ± 232.6	3	124.2 ± 18.4	10
16αOH-DHA	364.1 ± 74.7	3	146.7 ± 21.3	10
Estrone	8.4 ± 4.2	3	130.4 ± 28.3	9
Estradiol	6.8 ± 2.2	3	91.7 ± 21.4	9
Estriol	6.7 ± 0.2	3	154.8 ± 23.0	9
Estetrol	1.6 ± 0.3	3	6.5 ± 1.0	9

n = Number of cases.

Hormone Levels in Urine

Very low estriol excretion rates, i.e. values not exceeding 3 mg per day, were noted in all the reported cases. Urinary estrone and estradiol excretion rates were similarly depressed, but pregnanediol [4] and 17-ketosteroid [5] excretion rates were normal. Excretion of 11 urinary steroid monosulfates has recently been reported in 9 cases of placental sulfatase deficiency by *Taylor and Shackleton* [16]. These authors measured steroid sulfates by gas chromatography and found that the excretion of all steroids was very high except for DHA; excretion of 16α-OH-pregnenolone was 33-fold, androstenetriol was 14-fold, and 16α-OH-DHA was 19-fold in excess of mean normal values. The steroid sulfate of 16-oxo-androstenediol and 15β-, 16α-dihydroxy-DHA concentrations in urine seemed to be high.

Hormone Levels in Maternal Blood

Not only the maternal urinary estriol levels, but also the maternal plasma levels of estriol were low in pregnancies with placental sulfatase deficiency [7, 8, 12, 18]. Estrone [18], estradiol [3, 18] and estetrol levels [18] were also found to be low. *Osathanondh* et al. [8] reported that the level of maternal serum estetrol was undetectable, whereas DHA-S was normal. The plasma concentrations of estradiol, progesterone, 17αOH-progesterone, 11β-hydroxycorticosteroids were reported to be subnormal by *Oakey* et al. [5] in a woman with placental sulfatase deficiency. The free and conjugated steroid levels in maternal venous blood measured in 3 patients are shown in table I [18]. The levels of all estrogens were low compared with levels in normal pregnant women. The concentrations of androgens, especially 16α-OH-DHA-sulfate, were higher than normal. The conjugated Δ^5-steroid levels, including pregnenolone and 16αOH-pregnenolone, were higher than normal, whereas the progesterone and 20α-dihydroprogesterone levels were normal except in 1 case. It is noteworthy that the levels of 16α-hydroxylated Δ^4-steroids such as 16αOH-testosterone, 16αOH-androstenedione and 16αOH-progesterone were subnormal. Elevated maternal plasma levels of DHA-S and 16αOH-DHA-S were also reported by *Mango* et al. [14] and *Taylor* et al. [16] while another report [8] demonstrated that DHA-S levels were normal.

Beside steroid, serum concentrations of placental lactogen [3, 9, 19] were not affected in pregnancies with placental sulfatase deficiency.

Changes in the steroid levels in the maternal circulation may be instrumental in the prenatal diagnosis of placental sulfatase deficiency.

Hormone Levels in Cord Blood

Absence of sulfatase activity in the placenta may alter the steroidal profile in the feto-placento-maternal compartment. Placental sulfatase deficiency, therefore, may provide information on the mechanisms of steroid biosynthesis in the fetoplacental unit. Because of this biochemical aspect, steroid levels in cord blood have aroused several investigators' interest. Although some steroids have been reported to remain normal in cord blood, elevated levels of DHA sulfate were reported by *Osathanondh* et al. [8], and *France* et al. [4] demonstrated that cord plasma levels of 16αOH-DHA-sulfate and pregnenolone sulfatase were within the normal range. 12- and 18-fold increases in androstenetriol were noted in 2 patients with sulfatase deficiency, while the levels of DHA-sulfate, 16αOH-DHA-sulfates and 16-keto-androstenedione were comparable to those of normal pregnancies [6].

Table II. Steroid levels in venous cord blood (mean ± SE; ng/ml)

Steroids	Placental sulfatase deficiency		n	Normal		n
Free steroid						
Pregnenolone	274.8 ±	119.2	4	67.0 ±	3.8	9
Progesterone	120.9 ±	26.2	2	224.5 ±	20.1	9
16αOH-preg.	5.2 ±	0.1	2	10.4 ±	1.3	10
16αOH-prog.	6.7 ±	2.8	2	24.5 ±	4.3	10
20αOH-prog.	115.6		1	50.7 ±	6.1	10
DHA	5.7 ±	1.4	4	4.4 ±	0.4	10
16αOH-DHA	8.6 ±	2.1	4	23.2 ±	2.9	10
16αOH-A-dione	2.1 ±		1	10.1 ±	1.2	5
16αOH-test	2.2		1	6.7 ±	1.0	5
Estrone	2.4 ±	1.4	3	9.1 ±	1.4	10
Estradiol	1.0 ±	0.5	3	7.0 ±	1.1	8
Estriol	0.9 ±	0.2	3	43.7 ±	8.8	9
Estetrol	0.1		1	5.7 ±	0.8	10
Conjugated steroid						
Pregnenolone	2,035.8 ±	628.0	4	741.8 ± 117.3		10
16αOH-preg.	1,288.0 ±	277.0	2	572.3 ± 182.5		10
DHA	603.8 ±	145.2	4	484.7 ± 64.2		10
16αOH-DHA	17,416.0 ± 11,956.2		3	2,478.6 ± 346.5		9
Estrone	5.6 ±	1.5	3	44.1 ±	9.0	10
Estradiol	20.0 ±	7.2	3	96.1 ±	10.4	8
Estriol	22.5 ±	7.8	4	2,340.2 ± 438.3		10
Estetrol	3.7 ±	2.9	2	65.6 ±	10.0	10

n = Number of cases.

Mean cord blood concentrations of 6 Δ^5-steroid sulfates were demonstrated by *Taylor and Shackleton* [16] to be higher in 4 patients with placental sulfatase deficiency than in 4 normal subjects but the differences were not significant. Great interindividual variations in cord plasma steroid levels may account for these differences. Table II shows the steroid levels in cord blood measured by us. The levels of free and conjugated estrogens were extremely low compared with normal values while the levels of 16αOH-DHA-sulfate and 16αOH-pregnenolone sulfate were high. The levels of progesterone and 16αOH-progesterone were lower than normal. The finding of raised levels of steroid sulfates in cord plasma associated with decreased levels of Δ^4-ketosteroid may suggest the biosynthetic pathways of the steroid in the fetoplacental compartment.

Hormone Levels in Amniotic Fluid

In amniotic fluid from patients with placental sulfatase deficiency, *Osathanondh* et al. [8] and *Braunstein* et al. [7] recorded increased amounts of DHA-S in each case. *Taylor and Shackleton* [16] reported high mean levels of DHA-S, 16αOH-DHA, androstenetriol and 16αOH-pregnenolone in 3 patients; these values were in contrast to the values of pooled amniotic fluid from normal women, although the values were widely spread.

The concentration of pregnanetriol in the amniotic fluid reported by *Oakey* et al. [5] was much higher than the values observed in normal pregnancies.

Clinical Aspects

Diagnosis

The disorder is usually suspected when patients at term are found to have extremely low urinary estriol excretion without any evidence of fetal distress or anencephaly. Idiopathic congenital adrenal fetal hypoplasia can also be at the origin of low maternal urinary estriol excretion. Identification of each of these disorders would be of great value in the prognosis of the outcome of pregnancy.

Fetal death in utero and anencephaly can be easily diagnosed by routine clinical procedures, absence of fetal heart tones and radiography. A reliable method for distinguishing prenatally between fetal adrenal hypoplasia and placental sulfatase deficiency would be desirable, since the former disorder gravely endangers the neonate and the latter is often associated with difficulty in labor.

Osathanondh et al. [8] reported that the combination of extremely low estriol and estetrol concentrations in the maternal venous blood with elevated amniotic fluid DHA-S and normal amniotic fluid DHA appeared to be characteristic of placental sulfatase deficiency. Congenital adrenal hypoplasia and anencephaly can be ruled out since these abnormalities would be associated with subnormal levels of free and conjugated DHA in the amniotic fluid.

Prenatal diagnosis of placental sulfatase deficiency may be possible on the basis of DHA-S and/or 16αOH-DHA-sulfate levels in urine and/or blood [14]. *Taylor and Shackleton* [16] found high excretion of 16αOH-DHA-sulfate was characteristic of placental sulfatase deficiency.

Steroidal profiles of maternal blood may be useful for the diagnosis of

Table III. Mode of delivery in placental sulfatase deficiency

Mode of delivery	Cases n	%
Spontaneous labor; vaginal delivery	6	16.2
Spontaneous labor; cesarean section	5	13.5
Induction of labor; vaginal delivery	7	18.9
Induction of labor; cesarean section	11	29.8
Elective cesarean section	8	21.6
Total	37	100

sulfatase deficiency. *Beastall* et al. [9] stated that low urinary estriol combined with normal levels of HPL was useful for the screening of placental sulfatase deficiency.

Yanaihara et al. [18] have shown (table I) that high concentrations of maternal serum conjugated 16α-OH-Δ^5-steroids, including 16α-OH-DHA and 16α-OH-pregnenolone associated with low 16α-OH-progesterone and low estrogens, may help in the diagnosis of placental sulfatase deficiency.

A DHA and DHA-S loading test has been used to make a prenatal diagnosis of placental sulfatase deficiency. *France* et al. [4] infused these steroids into a patient with low urinary estriol excretion. DHA-S infusion did not result in a rise in urinary estrogens whereas an increase was observed in urinary estrogen excretion following DHA administration. Instillation of DHA-S into amniotic fluid was suggested by *Tabei and Heinrichs* [6]. They observed no rise of urinary estriol excretion in patients with placental sulfatase deficiency. We have demonstrated that maternal serum estrogens remained unchanged following administration of DHA-S into the peripheral circulation of a patient.

Mode of Parturition

Patients with placental sulfatase deficiency characteristically fail to develop spontaneous labor. In the past reports, high rates of cesarean section were recorded because of failure of induction of labor. Table III summarizes the modes of delivery reported in the published cases of placental sulfatase deficiency [1–18]. Pregnancy was terminated by cesarean section in 65% of the reported cases. It should be noted that spontaneous labor was observed in 16% of the cases. In the rest of the cases, the labor was induced. The occurrence of spontaneous deliveries does not agree with the concept that

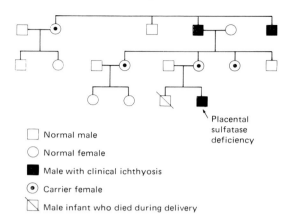

Fig. 3. Family pedigree of a case of placental sulfatase deficiency [case 2 in our study; 18, 19].

estrogen is required for cervical maturing and initiation of parturition [16]. Normal uterine responses of oxytocin or prostaglandin have been found in some of the patients with this disorder [6, 16, 18]. It has been noted that spontaneous labor is more frequent in multigravidae [16] and that primigravidae go beyond term.

Infants

All affected fetuses to date, except for the cases reported by *Mango* et al. [14], were males and their intrauterine growth was normal. The average birth weight of the infants of patients with placental sulfatase insufficiency is 3,057.3 g (n = 33) and the mean weight of the placentas is 523.6 g (n = 27). Histologically, those placentas were reported to be normal. Only 2 of 33 babies recorded were below 2,500 g. These 2 infants were delivered at 35 and 38 weeks of gestation, respectively. It appears that placental sulfatase insufficiency does not affect intrauterine growth. Although no abnormality was observed at birth, the infants developed ichthyosis during the period of 1–3 months after birth. 2 of our 4 cases developed ichthyosis.

Since almost all the infants were males, ichthyosis is considered to be X-linked. The relationship between placental sulfatase deficiency and ichthyosis was initially suggested by *Shapiro* et al. [17]. They observed 2 cases of placental sulfatase deficiency who were both males and developed ichthyosis. The infants belonged to two genealogically unrelated families both

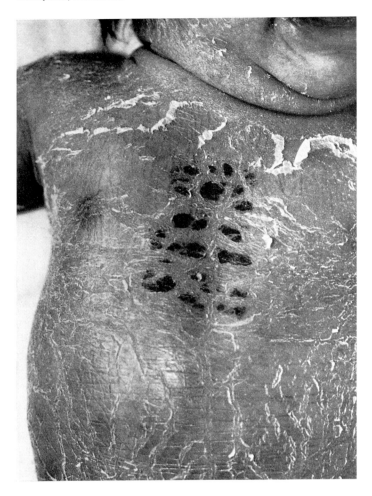

Fig. 4. Trunk of a male infant who developed ichthyosis at 4 months of age. The placenta was confirmed as having sulfatase deficiency (kindly supplied by Dr. *N. Nagamachi*, Kagawa, Japan).

of which had several male members affected with ichthyosis. *Shapiro* et al. [17] found that the sulfatase activities of fibroblast obtained from those ichthyosis-affected individuals were extremely low. Figure 3 shows the family pedigree of our case 2. Ichthyosis seems to be X-linked as it affects only males [19]. Figure 4 shows the ichthyosis which developed 4 months after delivery. Deficient placental sulfatase activity was confirmed in our laboratory. This case is not included in the above-mentioned 4 cases.

The incidence of placental sulfatase deficiency in pregnant females is unknown. *Oakey* et al. [5] reported that only 1 such case could be detected among 5,000 complicated pregnancies. *Taylor and Shackleton* [16] have recently reported that 6 cases had been recorded in a 15-month period from a yearly total of 4,700 pregnant females.

Acknowledgements

We are indebted to Drs. *Y. Fukushima*, *K. Takahashi* and *T. Tabei* for their kind help in supplying us with the clinical material, and Dr. *N. Nagamachi* for supplying us with the photograph of the ichthyosis-affected infant whose placenta we have recently confirmed to have placental sulfatase deficiency. We would like to express our thanks to Miss *K. Hirato* for the assay of the placental enzyme activities.

References

1 France, J.T.; Liggins, G.C.: Placental sulfatase deficiency. J. clin. Endocr. Metab. *29:* 138–141 (1969).
2 Cedard, L.; Tchobrousky, C.; Guglielmina, R.; Mailhac, M.: Insuffisance œstrogénique paradoxale au cours d'une grossesse normale par défaut de sulfatase placentaire. Bull. Féd. Socs Gynéc. Obstét. Lang. fr. *23:* 16–20 (1971).
3 Fliegmer, J.R.H.; Schindler, I.; Brown, J.B.: Low urinary oestriol excretion during pregnancy associated with placental sulphatase deficiency of congenital adrenal hypoplasia. J. Obstet. Gynaec. Br. Commonw. *79:* 810–815 (1972).
4 France, L.T.; Seddon, R.J.; Liggins, G.C.: A study of a pregnancy with low estrogen production due to placental sulfatase deficiency. J. clin. Endocr. Metab. *36:* 1–9 (1973).
5 Oakey, R.E.; Cawood, M.L.; Macdonald, R.R.: Biochemical and clinical observations in a pregnancy with placental sulphatase and other enzyme deficiencies. Clin. Endocr. *3:* 131–148 (1974).
6 Tabei, T.; Heinrichs, W.L.: Diagnosis of sulfatase deficiency. Am. J. Obstet. Gynec. *124:* 409–414 (1976).
7 Braunstein, G.D.; Ziel, F.H.; Allen, A.; Velde, R. Van de; Wade, M.E.: Prenatal diagnosis of placental steroid sulfatase deficiency. Am. J. Obstet. Gynec. *126:* 716–719 (1976).
8 Osathanondh, R.; Canick, J.; Ryan, K.J.; Tulchinsky, D.: Placental sulfatase deficiency: a case study. J. clin. Endocr. Metab. *43:* 208–214 (1976).
9 Beastall, G.H.; Kelly, A.M.; England, P.; Rao, L.G.S.; MacGregor, M.W.; Paterson, M.L.: Urinary oestrogen and plasma human placental lactogen as initial screening test for a placental sulphatase deficiency. Scot. med. J. *21:* 106–108 (1976).

10 Bedin, M.; Conquy, P.; Alsat, E.; Cedard, L.: Déficit en sulfatase placentaire. Nouv. Presse méd. *5:* 1889–1892 (1976).
11 Chadwick, J.M.; Murnain, J.R.: Placental sulphatase deficiency causing low urinary oestriol excretion in pregnancy complicated by toxesmia. Aust. N.Z. J. Obstet. Gynaec. *16:* 119–120 (1976).
12 Shapiro, L.J.; Cousins, L.; Fluharty, A.V.; Stevens, R.I.; Kihara, H.: Steroid sulfatase deficiency. Pediat. Res. *11:* 894–897 (1977).
13 Lehmann, W.D.; Lauritzen, C.: Clinical and biochemical studies in three pregnancies with placental sulfatase deficiency. Acta endocr., Copenh., suppl. 215, p. 36 (1978).
14 Mango, D.; Montemurro, A.; Scirpa, P.; Bompiani, A.; Menini, E.: Four cases of pregnancy with low estrogen production due to placental enzymatic deficiency. Eur. J. Obstet. Gynecol. reprod. Biol. *8:* 65–71 (1978).
15 Yanaihara, T.; Takayama, T.; Saitoh, H.; Hirato, K.; Nakayama, T.; Fukushima, Y.; Horiguchi, M.; Ishihara, T.; Takahashi, K.; Tabei, T.: Two cases of placental sulfatase deficiency. Acta obstet. gynec. jap. *31:* 361–362 (1979).
16 Taylor, N.F.; Shackleton, C.H.L.: Gas chromatographic steroid analysis for diagnosis of placental sulfatase deficiency: a study of nine patients. J. clin. Endocr. Metab. *49:* 78–86 (1979).
17 Shapiro, L.J.; Weiss, R.; Webster, D.; France, J.T.: X-Linked ichthyosis due to steroid-sulfatase deficiency. Lancet *1:* 70–72 (1978).
18 Yanaihara, T.; Hirato, K.; Saitoh, H.; Hashino, M.; Kojima, S.; Nakayama, T.: Four cases of placental steroid sulphatase deficiency: steroid levels in maternal vein blood. J. Foet. Med. *1:* 49–53 (1981).
19 Fukushima, Y.; Horiguchi, M.; Ishihara, T.; Yanaihara, T.; Hirato, K.; Nakayama, T.: A case of placental sulfatase deficiency: in relation to X-linked ichthyosis. Acta obstet. gynaec. jap. *33:* 420–423 (1981).

Prof. T. Nakayama, MD, Department of Obstetrics and Gynecology, School of Medicine, Showa University, Tokyo (Japan)

New Placental Proteins in Placental Dysfunction

J.N. Lee, T. Chard

Departments of Reproductive Physiology and Obstetrics and Gynaecology, St. Bartholomew's Hospital Medical College, and The London Hospital Medical College, London, UK

Introduction

The human placenta secretes a variety of proteins into the maternal circulation. These include hormones, enzymes, and a series of newly discovered proteins with no obvious function (tables I and II). Measurement of these placental products has been widely used as an indicator of placental function and, in turn, as a diagnostic marker of the well-being of the fetus.

The New Placental Proteins

The characteristics of the new placental proteins are summarised in table II. All appear to be synthesised by the syncytiotrophoblast [*Bohn and Sedlacek*, 1975; *Lin and Halbert*, 1976; *Horne* et al., 1976b; *Inaba* et al., 1981], though there is some evidence for a decidual origin of 'Schwangerschaftsprotein 1' (SP1) [*Koh and Cauch*, 1981] and pregnancy-associated plasma protein A (PAPP-A) [*Bischof* et al., 1982]. The control mechanisms for the production of these proteins is poorly understood, and it has been suggested that their circulating levels in the mother are related solely to trophoblast mass and uteroplacental blood flow [*Gordon and Chard*, 1979]. The ratio of serum to placental concentrations differs widely amongst the proteins, suggesting that there is no single process of secretion [*Lee* et al., 1979].

Functions of the New Placental Proteins

A number of possible functions have been suggested for SP1: for example, as a factor in the control of carbohydrate metabolism [*Tatra* et al., 1976]; as a carrier of steroid hormones and iron [*Bohn and Kranz*, 1973;

Table I. Protein hormones (and synonyms) and enzymes produced by the human placenta

	Abbreviations
1. Proteins	
Human chorionic gonadotrophin	hCG
Human placental lactogen	hPL
(Human chorionic somatomammotrophin)	(hCS)
2. Peptides	
Human chorionic luteinising hormone-releasing hormone	hC-LHRH
Human chorionic thyrotrophin-releasing hormone	hC-TRH
Human chorionic growth hormone-releasing inhibiting hormone	–
(Human chorionic somatostatin)	–
Human chorionic adrenocorticotrophin and its related peptide	–
Human chorionic lipotrophin and its related peptide	–
3. Enzymes	
Heat-stable alkaline phosphatase	HSAP
Cystine aminopeptidase (oxytocinase)	CAP
Diamine oxidase (histamine)	DAO

Lin et al., 1974a, b], and as an immunosuppressive agent [*Cerni* et al., 1977]. However, none of these functions seems essential to pregnancy, since an apparent absence of SP1 may be associated with a normal outcome [*Grudzinskas* et al., 1979a]. Recent evidence suggests that placental protein 5 (PP5) may be related to the coagulation system during pregnancy [*Salem* et al., 1981b, c], and that PAPP-A may act as an inhibitor of the complement system and thus have an immunosuppressive function [*Bischof*, 1981].

Clinical Application of the New Placental Proteins

SP1

SP1 in Normal Pregnancy. SP1 can be detected in maternal blood 18–23 days after the LH surge [*Grudzinskas* et al., 1977a, b]. Thereafter there is an exponential rise, the levels doubling every 2–3 days; the slope decreases at 40 days [*Grudzinskas* et al., 1979a, b]. The levels continue to rise to reach a plateau at 36 weeks with a mean value variously estimated at 95–250 mg/l [*Tatra* et al., 1974; *Towler* et al., 1976; *Gordon* et al., 1977a, b; *Klopper* et al., 1978; *Sorensen*, 1978; *Heikinheimo* et al., 1978; *Lee* et al., 1980b]. Levels are 1,000-fold less in fetal blood than in maternal blood, and 100-fold less

Table II. Characteristics of the new placental proteins

Proteins (and synonyms)	Abbreviations	Molecular weight daltons	Carbohydrate content %	Half-life
Schwangerschaftsprotein 1	SP1	90,000–430,000	29.3	17–45 h
(Pregnancy specific β_1-glycoprotein)	(PSβ_1G)			
(Trophoblast β_1-glycoprotein)	(TBG)			
(Trophoblast specific β-glycoprotein)	(TSG)			
(Pregnancy-associated plasma protein C (β_1-SP1)	(PAPP-C)			
Pregnancy-associated plasma protein A	PAPP-A	750,000	19.2	3–4 days
Pregnancy-associated plasma protein B	PAPP-B	1,000,000		< 1 day
Placental protein 5	PP5	42,000	19.8	15–30 min

in amniotic fluid [*Grudzinskas* et al., 1978]. Small and variable amounts are detected in urine.

SP1 in Early Pregnancy. Measurements of SP1 can be used for the detection of early pregnancy. In this respect it is equivalent to, or perhaps marginally less sensitive than assays for the β-subunit of human chorionic gonadotrophin (hCG) [*Grudzinskas* et al., 1979b]. Since nonpregnant subjects occasionally have detectable amounts of SP1 [*Searle* et al., 1978; *Wurz*, 1979; *Anthony* et al., 1980], the postulated advantage of SP1 measurement in terms of specificity has not been realised in practice.

SP1 in Abnormal Early Pregnancy. It has been proposed that measurement of SP1, like hCG, may detect 'occult' pregnancies, i.e. those which abort around the time of the first missed menstrual period [*Seppälä* et al., 1978a, c]. Measurement of both SP1 and hCG may be of value in the differential diagnosis of ectopic pregnancy [*Seppälä* et al., 1980; *Ouyang* et al., 1980; *Tatra* et al., 1981].

Low SP1 levels are an indicator of unfavourable outcome in threatened abortion [*Schultz-Larsen and Hertz*, 1978; *Jandial* et al., 1978; *Jouppila* et al., 1980], findings similar to those with both hCG [*Dhont* et al., 1975] and

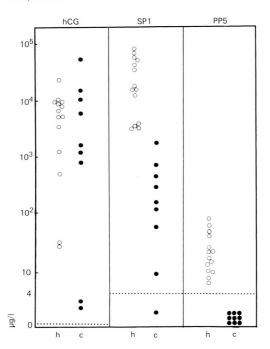

Fig. 1. Levels of hCG, SP1 and PP5 in untreated gestational trophoblastic tumours. ○ = Hydatidiform mole; ● = choriocarcinoma [from *Lee* et al., 1981b].

human placental lactogen (hPL) [*Niven* et al., 1972]. Decreased levels of SP1 are associated with anembryonic pregnancy [*Bennett* et al., 1978], and after prostaglandin-induced therapeutic abortion in the first trimester [*Mandelin* et al., 1978].

SP1 has been detected in vesicle fluid, tissue extracts, tissue sections and maternal blood and urine from cases of hydatidiform mole [*Searle* et al., 1978; *Seppälä* et al., 1978b; *Tatarinov* et al., 1974, 1976; *Horne* et al., 1977] and its measurement has been put forward as a useful prognostic indicator, in combination with hCG, in the management of patients with trophoblastic tumours [*Seppälä* et al., 1978c; *Lee* et al., 1982a]. Moreover, *Lee* et al. [1981b] reported that measurement of circulating levels of SP1 and PP5, but not hCG, provides a distinction between benign and malignant gestational trophoblastic tumours (fig. 1). Ectopic production of SP1 has also been noted in some non-trophoblastic tumours [*Horne* et al., 1976a; *Tatarinov and Sokolov*, 1977].

SP1 in Abnormal Late Pregnancy. Good information is now available on the value of SP1 in the clinical management of late pregnancy. It is generally accepted that decreased levels of SP1 are associated with intrauterine growth retardation (IUGR) both in the presence and absence of maternal hypertension [*Tatra* et al., 1975; *Chapman and Jones,* 1978; *Gordon* et al., 1977a; *Grudzinskas* et al., 1977a; *Lin* et al., 1974a, b; *Sorensen,* 1978; *Towler* et al., 1977; *Lee* et al., 1980b; *Viala* et al., 1980; *Wurz* et al., 1981]. It has been demonstrated that smoking affects fetus and placenta as a single unit in late pregnancy and decreases SP1 levels, but not hPL, in maternal circulation [*Lee* et al., 1980b, 1981a]. However, *Hughes* et al. [1980a, b] failed to find a significant difference in SP1 levels between mothers with growth-retarded infants and those with normal infants.

Levels are normal in hypertension/pre-eclampsia [*Chapman and Jones,* 1978; *Sorensen,* 1978; *Lin* et al., 1977; *Pluta* et al., 1979] unless there are fetal complications [*Tatra* et al., 1974] such as IUGR [*Siekmann and Heilmann,* 1979; *Towler* et al., 1977; *Wurz* et al., 1981]. Levels of SP1 are elevated in multiple pregnancy [*Lin* et al., 1977; *Tatra* et al., 1974; *Towler* et al., 1976; *Huang* et al., 1980].

Tatra et al. [1976, 1979] found increased levels of SP1 in amniotic fluid and maternal urine in association with excessive growth of the placenta. The levels are usually within the normal range in diabetic pregnancy [*Tatra* et al., 1974; *Grudzinskas* et al., 1979a; *Siekmann and Heilmann,* 1979] and are elevated in cases of cholestasis [*Heikinheimo* et al., 1978].

PP5

PP5 in Normal Pregnancy. Since maternal blood levels of PP5 are relatively low, measurement has only recently been possible using a specific radioimmunoassay [*Obiekwe* et al., 1979]. Circulating PP5 can be detected from the 8th week of gestation [*Grudzinskas* et al., 1979c; *Obiekwe* et al., 1979, 1980a; *Nisbet* et al., 1981] and thereafter rises to reach a plateau at 36 weeks with a median level of 32 μg/l [*Obiekwe* et al., 1979] which is approximately 100-fold less than that of hPL at the same time. After delivery it falls rapidly with a half-life of 5–39 min [*Obiekwe* et al., 1980b], or 5–14 min [*Nisbet* et al., 1981].

PP5 in Abnormal Pregnancy. Because of the low normal levels there is little information about the clinical use of PP5 in early pregnancy. However, the presence of PP5 has been reported in homogenates of trophoblastic tumours, in testicular teratomas, in mole vesicle fluid and in the serum of

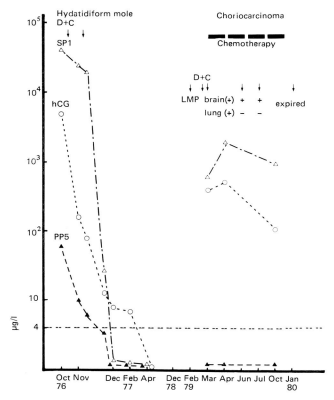

Fig. 2. Serum hCG, SP1 and PP5 levels in a 20-year-old patient with choriocarcinoma, 2 years after evacuation of a hydatidiform mole [PP5 was undetectable in sample collected during treatment of the choriocarcinoma; from *Lee* et al., 1982b].

some patients with hydatidiform mole [*Grudzinskas* et al., 1979b; *Lee* et al., 1982c]. Since PP5 is present in the blood of patients with hydatidiform mole but not in choriocarcinoma it has been suggested that PP5 measurements may be valuable in distinguishing between benign and malignant trophoblastic tumours [*Lee* et al., 1981b] (fig. 1). Furthermore, a study on 228 samples from 31 patients concluded that circulating PP5 levels relate closely to the invasive activity of neoplastic trophoblast [*Lee* et al., 1982b] (fig. 2).

In late pregnancy, initial studies suggested that PP5 assays were unpromising as a test for IUGR [*Obiekwe* et al., 1980a]. However, *Salem* et al. [1981a, 1982] found a significant elevation of PP5 in association with premature delivery and placental abruption. Furthermore, *Lee* et al. [1981c]

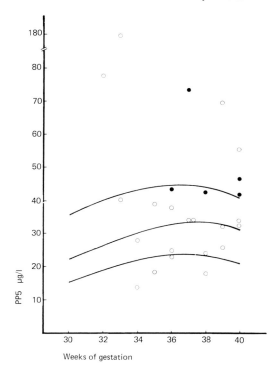

Fig. 3. Levels of serum PP5 in 20 patients with severe pre-eclampsia (○), and 5 patients with eclampsia (●). Solid lines indicate the 10th, median and 90th centiles of the normal range [from *Lee* et al., 1981c].

reported that PP5 levels were elevated in severe pre-eclampsia and eclampsia (fig. 3) and in severe rhesus isoimmunisation [*Lee* et al., 1982a].

PAPP-A

PAPP-A in Normal Pregnancy. Although PAPP-A can be measured by crossed immunoelectrophoresis [*Lin* et al., 1974a] and rocket immunoelectrophoresis [*Bischof* et al., 1979; *Folkersen* et al., 1981], the sensitivity of these techniques is inadequate for the estimation of circulating PAPP-A in pregnancy before 20 weeks of gestation. However, the recent introduction of a radioimmunoassay for PAPP-A [*Sinosich* et al., 1981] has permitted the detection of PAPP-A 28 days after conception. The level rises slowly to 30 weeks and thereafter rises more steeply up to term with a mean value variously estimated at 97–168 units per millilitre of plasma or 36.07–388.68

μg/ml [*Lin* et al., 1977; *Klopper* et al., 1979; *Smith* et al., 1979; *Hughes* et al., 1980a; *Bischof* et al., 1980, 1981; *Sinosich* et al., 1982]. The gross discrepancy between the values reported by different authors is probably due to differences in standards and in the specificity of the antibodies.

PAPP-A in Abnormal Pregnancy. Information on circulating PAPP-A in abnormal pregnancy is very limited. In early pregnancy, *Sinosich* et al. [1982] have reported that PAPP-A is detectable in patients with hydatidiform mole but not those with choriocarcinoma, a finding similar to that with PP5.

Levels of PAPP-A are raised in twin pregnancy [*Lin* et al., 1977; *Sinosich* et al., 1982]. In pre-eclampsia, levels may be raised though they show no relation to the presence or absence of fetal growth retardation. The highest levels are associated with severe cases including those with proteinuria [*Hughes* et al., 1980a, b; *Klopper and Hughes*, 1980; *Klopper*, 1980, 1981a, b]. The rise in PAPP-A levels may often precede the onset of the clinical condition [*Toop and Klopper*, 1981]. There is also some evidence that, like PP5, levels of PAPP-A may be elevated prior to the onset of placental abruption and premature labour [*Hughes* et al., 1980b].

Conclusions: The New Placental Proteins
in Relation to Placental Dysfunction and Fetal Risk

The present clinical findings with measurement of the new placental proteins are summarised in table III.

Current evidence suggests that maternal blood levels of all placental proteins are primarily a function of trophoblast mass and uteroplacental blood flow: there is no evidence for any direct relationship to the fetus. However, it is well-recognised that most complications of pregnancy (with the obvious exception of congenital abnormalities) have their origin in the placenta rather than the fetus itself and it is not surprising, therefore, that measurement of exclusively placental products provides a reasonably accurate overall reflection of fetal well-being. Equally, it must be emphasised that the 'accuracy' is relative rather than absolute [*Chard*, 1977]: there is always substantial overlap between normal and abnormal, and hence a frequent occurrence of false-positive and false-negative results. This problem is not unique to placental biochemistry, and applies with equal force to most parameters measured as part of antenatal care.

Table III. Circulating levels of the new placental proteins in relation to placental dysfunction and fetal risk

Clinical observation	Placental proteins		
	SP1	PP5	PAPP-A
Twin pregnancy	↑	↑	↑
Ectopic pregnancy	↓	–	–
Threatened abortion	↓	↓	–
Anembryonic pregnancy	↓	–	–
Molar pregnancy	↑	↑	↑
Diabetic pregnancy	N	↑	–
Pregnancy with Rhesus isoimmunisation	N	↑	–
Placental abruption	N	↑	↑
Premature delivery	N	↑	↑
Pre-eclampsia (PET) and eclampsia	↓	↑	↑
Intrauterine growth retardation	↓	–	–
With PET	–	↑	–
Without PET	N	↓	–
Trophoblastic tumours	↑	↑	–

N = Normal levels; ↑ = increased levels; ↓ = decreased levels.

Table IV. A new classification of placental products

	Group I	Group II
Products	enzymes, steroids, hCG, hPL, SP1	PP5, PAPP-A
Functions	hormonal[2]	local (immune and coagulation systems)
Pathology[1]	fetus[3]	placenta[4]
Clinical manifestations	IUGR, acute fetal distress	placental abruption, premature labour, pre-eclampsia

[1] The pathological changes refer to late pregnancy; in early pregnancy measurement of all fetoplacental products appears to give virtually identical results.
[2] 'Hormonal' is used here in the broad sense of action at a distance.
[3] Group I compounds reflect fetal well-being independently of other complications.
[4] Group II compounds reflect specific events which may lead to fetal risk, but which are associated with a normal fetus at their onset.

Traditionally, it is low levels of a placental product which are considered to be the abnormal, unfavourable sign. However, recent studies have shown that *elevated* levels are the characteristic pathological change of PP5 and PAPP-A. On the basis of this, together with certain biological observations, it has been proposed that placental products can be divided into two groups (table IV). The group I products (including SP1) seem to reflect the overall function of the placenta, while group II products (PP5 and PAPP-A) seem to reflect specific pathological events – most notably, those underlying the occurrence of pre-eclampsia, placental abruption, and premature labour. The unique predictive value of PP5 and PAPP-A in these conditions holds great promise for the future of this range of biochemical tests.

References

Anthony, F.; Masson, G.; Wood, P.: Br. J. Obstet. Gynaec. *87:* 496 (1980).
Bennett, M.J.; Grudzinskas, J.G.; Gordon, Y.B.; Turnbull, A.C.: Br. J. Obstet. Gynaec. *85:* 348 (1978).
Bischof, P.: Placenta *2:* 29 (1981).
Bischof, P.; Bruce, D.; Cunningham, P.; Klopper, A.: Clinica chim. Acta *95:* 243 (1979).
Bischof, P.; Du Berg, S.; Schindler, A.: In Klopper, The immunology of the placental proteins (Saunders, London, in press, 1982).
Bischof, P.; Haenggeli, L.; Sizonenko, M.; Herrmann, W.; Sizonenko, P.: Biol. Reprod. *24:* 1076 (1981).
Bischof, P.; Hughes, G.; Klopper, A.: Am. J. Obstet. Gynec. *138:* 494 (1980).
Bohn, H.; Kranz, T.: Arch. Gynaek. *215:* 63 (1973).
Bohn, H.; Sedlacek, H.: Arch. Gynaek. *220:* 105 (1975).
Cerni, C.; Tatra, G.; Bohn, H.: Arch. Gynaek. *223:* 1 (1977).
Chapman, M.G.; Jones, W.R.: Aust. N.Z. J. Obstet. Gynaec. *18:* 172 (1978).
Chard, T.: In Stallworthy, Bourne, Recent advances in obstetrics and gynaecology, vol. 12, p. 146 (Churchill, Livingstone, Edinburgh 1977).
Dhont, M.; Thiery, M.; Vanderkerckhove, D.; Van Cauwenberghe, A.: Tijdschr. Geneesk. *22:* 1097 (1975).
Folkersen, J.; Grudzinskas, J.G.; Hindersson, P.; Teisner, B.; Westergaard, J.: Am. J. Obstet. Gynec. *139:* 910 (1981).
Gordon, Y.B.; Chard, T.: In Klopper, Chard, Placental proteins, p. 1 (Springer, Heidelberg 1979).
Gordon, Y.B.; Grudzinskas, J.G.; Jeffrey, D.; Chard, T.; Letchworth, A.T.: Lancet *i:* 331 (1977a).
Gordon, Y.B.; Grudzinskas, J.G.; Lewis, J.D.; Jeffrey, D.; Letchworth, A.T.: Br. J. Obstet. Gynaec. *84:* 642 (1977b).

Grudzinskas, J.G.; Evans, D.G.; Gordon, Y.B.; Jeffrey, D.; Chard, T.: Obstet. Gynec., N.Y. 52: 43 (1978).
Grudzinskas, J.G.; Gordon, Y.B.; Jeffrey, D.J.; Chard, T.: Lancet i: 333 (1977a).
Grudzinskas, J.G.; Humphreys, J.D.; Brudenell, M.; Chard, T.: Br. J. Obstet. Gynaec. 86: 978 (1979a).
Grudzinskas, J.G.; Lendon, E.A.; Gordon, Y.B.; Kelso, I.M.; Jeffrey, D.; Sobowale, O.; Chard, T.: Br. J. Obstet. Gynaec. 84: 740 (1977b).
Grudzinskas, J.G.; Lendon, E.A.; Obiekwe, B.C.: In Klopper, Chard, Placental proteins, p. 119 (Springer, Heidelberg 1979b).
Grudzinskas, J.G.; Menabawey, M.; Wiley, B.A.; Teisner, B.; Chard, T.: Br. J. Obstet. Gynaec. 86: 891 (1979c).
Heikenheimo, M.; Unnerus, H.A.; Ranta, T.; Jalanko, H.; Seppälä, M.: Obstet. Gynec., N.Y. 52: 276 (1978).
Horne, C.H.W.; Reid, I.N.; Milne, G.D.: Lancet ii: 279 (1976a).
Horne, C.H.W.; Towler, C.M.; Pugh-Humphreys, R.G.P.; Thomson, A.W.; Bohn, H.: Experientia 32: 1197 (1976b).
Horne, C.H.W.; Towler, C.M.; Milne, G.D.: J. clin. Path. 30: 19 (1977).
Huang, S.C.; Lee, J.N.; Chard, T.; Ouyang, P.C.; Wei, P.Y.: Int. J. biol. Res. Preg. 1: 159 (1980).
Hughes, G.; Bischof, P.; Wilson, G.; Klopper, A.: Br. med. J. i: 671 (1980a).
Hughes, G.; Bischof, P.; Wilson, G.; Smith, R.; Klopper, A.: Br. J. Obstet. Gynaec. 87: 650 (1980b).
Inaba, N.; Renk, T.; Weinmann, E.; Bohn, H.: Arch. Gynaek. 230: 195 (1981).
Jandial, V.; Towler, C.M.; Horne, C.H.W.; Abramovich, D.R.: Br. J. Obstet. Gynaec. 85: 832 (1978).
Jouppila, P.; Seppälä, M.; Chard, T.: Lancet i: 667 (1980).
Klopper, A.: In Sakamoto, Tojo, Nakayama, Gynecology and obstetrics. Excerpta Med. Int. Congr. Ser., No. 512, p. 1066 (1980).
Klopper, A.: In Kurjak, Rippman, Salovic, Current status of EPH gestosis, p. 288 (Excerpta Medica, Amsterdam 1981a).
Klopper, A.: In Givens, Endocrinology of pregnancy, p. 203 (Year Book Medical Publishers, Chicago 1981b).
Klopper, A.; Buchan, P.; Wilson, G.: Br. J. Obstet. Gynaec. 85: 738 (1978).
Klopper, A.; Hughes, G.: In Klopper, Genazzani, Crosignani, The human placenta: proteins and hormones, p. 17 (Academic Press, London 1980).
Klopper, A.; Smith, R.; Davidson, I.: In Klopper, Chard, Placental proteins (Springer, Heidelberg 1979).
Koh, S.H.; Cauch, M.: Eur. J. Obstet. Gynaec. Reprod. Biol. 11: 215 (1981).
Lee, J.N.; Grudzinskas, J.G.; Chard, T.: Br. J. Obstet. Gynaec. 86: 888 (1979).
Lee, J.N.; Grudzinskas, J.G.; Chard, T.: J. Obstet. Gynaec. 1: 87 (1980a).
Lee, J.N.; Grudzinskas, J.G.; Chard, T.: Obstet. Gynec., N.Y. 57: 220 (1981a).
Lee, J.N.; Huang, S.C.; Ouyang, P.C.; Wei, P.Y.; Al-Ani, A.T.M.; Chard, T.: J. Formosan med. Ass. 97: 986 (1980b).
Lee, J.N.; Lee, J.H.; Huang, S.C.; Ouyang, P.C.; Chard, T.: Submitted for publication (1982a).
Lee, J.N.; Salem, H.T.; Chard, T.; Huang, S.C.; Ouyang, P.C.: Br. J. Obstet. Gynaec. 89: 69 (1982b).

Lee, J.N.; Salem, H.T.; Al-Ani, A.T.M.; Chard, T.; Huang, S.C.; Ouyang, P.C.; Wei, P.Y.; Seppälä, M.: Am. J. Obstet. Gynec. *139:* 702 (1981b).

Lee, J.N.; Salem, H.T.; Huang, S.C.; Ouyang, P.C.; Seppälä, M.; Chard, T.: Int. J. Gynaec. Obstet. *19:* 65 (1981c).

Lee, J.N.; Wahlström, T.; Seppälä, M.; Salem, H.T.; Ouyang, P.C.; Chard, T.: Placenta *3:* 67 (1982c).

Lin, T.M.; Halbert, S.P.: Science *193:* 1249 (1976).

Lin, T.M.; Halbert, S.P.; Kiefer, D.; Spellacy, W.N.; Gall, S.: Am. J. Obstet. Gynec. *118:* 223 (1974a).

Lin, T.M.; Halbert, S.P.; Spellacy, W.N.: J. clin. Invest. *54:* 576 (1974b).

Lin, T.M.; Halbert, S.P.; Spellacy, W.N.; Berne, B.H.: Am. J. Obstet. Gynec. *128:* 808 (1977).

Mandelin, M.; Rutanen, E.M.; Heikenheimo, M.; Jalanko, H.; Seppälä, M.: Obstet. Gynec., N.Y. *52:* 314 (1978).

Nisbet, A.D.; Bremner, R.D.; Herriot, R.; Jandial, V.; Horne, C.H.; Bohn, H.: Br. J. Obstet. Gynaec. *88:* 484 (1981).

Niven, P.A.R.; Landon, J.; Chard, T.: Br. med. J. *ii:* 799 (1972).

Obiekwe, B.; Gordon, Y.B.; Grudzinskas, J.G.; Chard, T.; Bohn, H.: Clinica chim. Acta *95:* 509 (1979).

Obiekwe, B.; Grudzinskas, J.G.; Chard, T.: Br. J. Obstet. Gynaec. *87:* 302 (1980a).

Obiekwe, B.; Grudzinskas, J.G.; Chard, T.: In Klopper, Genazzani, Crosignani, The human placenta: proteins and hormones, p. 129 (Academic Press, London 1980b).

Ouyang, P.C.; Huang, S.C.; Lee, J.N.; Chard, T.; Wei, P.Y.: Int. J. biol. Res. Preg. *1:* 179 (1980).

Pluta, M.; Hardt, W.; Schmidt-Gollwitzer, K.; Schmidt-Gollwitzer, M.: Arch. Gynaek. *227:* 327 (1979).

Salem, H.T.; Lee, J.N.; Seppälä, M.; Vaara, L.; Aula, P.; Al-Ani, A.T.M.; Chard, T.: Br. J. Obstet. Gynaec. *88:* 371 (1981a).

Salem, H.T.; Seppälä, M.; Chard, T.: Placenta *2:* 205 (1981b).

Salem, H.T.; Seppälä, M.; Ranta, T.; Bohn, H.; Chard, T.: Br. J. Obstet. Gynaec. *88:* 367 (1981c).

Salem, H.T.; Westergaard, J.G.; Hindersson, P.; Seppälä, M.; Chard, T.: Br. J. Obstet. Gynaec. (in press, 1982).

Schultz-Larsen, P.; Hertz, J.B.: Eur. J. Obstet. Gynaec. reprod. Biol. *8:* 5 (1978).

Searle, F.; Leake, B.A.; Bagshawe, K.D.; Dent, J.: Lancet *i:* 579 (1978).

Seppälä, M.; Lehtovirta, P.; Rutanen, E.M.: Acta endocr., Copenh. *88:* 164 (1978).

Seppälä, M.; Rutanen, E.M.; Heikenheimo, M.; Jalanko, H.; Engvall, E.: Int. J. Cancer *21:* 265 (1978b).

Seppälä, M.; Rutanen, E.M.; Jalanko, H.; Lehtovirta, P.; Stenman, U.G.; Engvall, E.: J. clin. Endocr. Metab. *47:* 1216 (1978c).

Seppälä, M.; Venesmaa, P.; Rutanen, E.: Am. J. Obstet. Gynec. *136:* 189 (1980).

Siekmann, U.; Heilmann, L.: Z. Geburtsh. Perinat. *183:* 351 (1979).

Sinosich, M.J.; Teisner, B.; Davey, M.; Grudzinskas, J.G.: Aust. N.Z. J. Med. *11:* 429 (1981).

Sinosich, M.J.; Teisner, B.; Folkersen, J.; Saunders, D.M.; Grudzinskas, J.G.: Clin. Chem. *28:* 50 (1982).

Smith, R.; Bischof, P.; Hughes, G.; Klopper, A.: Br. J. Obstet. Gynaec. *86:* 882 (1979).
Sorensen, S.: Acta obstet. gynec. scand. *57:* 193 (1978).
Tatarinov, V.; Mesnyankina, N.V.; Nikovlina, D.M.; Novikora, L.A.; Toloknov, B.O.; Falaleeva, D.M.: Int. J. Cancer *14:* 548 (1974).
Tatarinov, Y.S.; Sokolov, A.V.: Int. J. Cancer *19:* 161 (1977).
Tatra, G.; Breitenecker, G.; Gruber, W.: Arch. Gynaek. *217:* 383 (1974).
Tatra, G.; Placheta, P.; Breitenecker, G.: Wien. klin. Wschr. *87:* 279 (1975).
Tatra, G.; Polak, S.; Placheta, P.: Arch. Gynaek. *221:* 161 (1976).
Tatra, G.; Polak, S.; Nasr, F.: Geburtsh. Frauenheilk. *41:* 359 (1981).
Tatra, G.; Riss, P.; Bohn, H.: Z. Geburtsh. Perinat. *183:* 254 (1979).
Toop, K.; Klopper, A.: In Miller, Thiede, Placenta: receptors, pathology and toxicology, p. 167 (Saunders, London 1981).
Towler, C.M.; Horne, C.H.W.; Jandial, V.; Campbell, D.M.; MacGillivray, I.: Br. J. Obstet. Gynaec. *83:* 775 (1976).
Towler, C.M.; Horne, C.H.W.; Jandial, V.; Campbell, D.M.; MacGillivray, I.: Br. J. Obstet. Gynaec. *84:* 258 (1977).
Viala, J.; Bastide, J.; Guibal, J.; Oudghiri, T.; Donati, R.: J. Gynec. Obstet. Biol. Reprod. *9:* 419 (1980).
Wurz, H.: Arch. Gynaek. *227:* 1 (1979).
Wurz, H.; Geiger, W.; Künzig, H.J.; Jabs-Lehmann, A.; Bohn, H.; Lüben, G.: J. perinatal Med. *9:* 67 (1981).

J.N. Lee, MB, MD, PhD, Department of Obstetrics and Gynecology,
Kaohsiung Medical College, Kaohsiung (Taiwan)

Subject Index

Abnormal depth of implantation 4
Abortion 41, 42
Acardiacus amorphus fetuses 10
Acute atherosis 51, 59, 62, 63, 67, 93–96
– –, affected vessel 93
– – with fibrinoid wall necrosis 64
– degenerative arteriolitis 92
– – changes, media 94
– fetofetal transfusion 10
AD/BPD 136
AD-measurements 136
Alcohol 6
Amniocentesis, hazards 6
Amnion nodosum 3, 10
Amnionitis 122
Amniotic band syndrome 14
– fluid 114, 161
– – DHA-S, elevated 151
– – , hormone levels 151
– – infection syndrome 115
– placental smears 122
Amyloid 49
Anaerobic bacteria 123
Androgens 149
Androstenedione 146
Androstenetriol 148
Anencephaly 145, 151
Aneuploidy 5
Aneurysms, chorionic vessels 13
Antigenicity, trophoblast 42
Aromatizing enzyme activity 146
Arterial wall 106

Arteries, placental bed 51
Arteriosclerosis 62
Arteriovenous shunts 11
Artery walls 96
Ascending infection 113
– –, amniotic cavity 12
– intrauterine infection 115, 118
At-risk group, size 130
Autologous antigens 43
Avascular terminal villi 111
– villi 106
Average fetal weight ratio 59
– placental weight, pure type 59
– – – ratio 59

Bacterial chorioamnionitis 109
Basal calcification 34
– decidua 86
– plate 86
Bilobed placenta 4
Biparietal diameters (BPD) 133
Blocking activity 42
– antibodies, formation 43
Blood group incompatibility 41
Borderline hypertension 96
BPD 133
– measurement 134, 135

Cardiac tumors 11
Calcification 37
Capillary basal lamina 81
– surface densities 23

Subject Index

Cellular hyperreactivity 46
– –, mother 52
Cesarean section 152
Chemotactic infectious agents 114
Chlamydia 123
Cholestasis, pregnancy 8
Chorangiomas 13
Chorangiosis 9
Chorioamnionitis 114, 115
Chorioangioma 61
Chorioangiomatosis diffusa 108
Choriocarcinoma 13, 41, 43, 162
Chorionic placental smears 122
– plate 33–35
– villi 32, 68
– –, hematogenous colonization 12
Chromosomal aberrations (trysomy 13, 18, 21, XYY syndrome) 45
– aneuploidy 4
– errors 7
– polyploidy 4
Chronic essential hypertension 20
– hypertension 21, 96
– intervillous hypoxia 7
– intrauterine inflammation 12
– nephritis 59
Circulating levels, new placental proteins 165
Circulatory disturbances 110
Circumvallate placenta 4
Circumvallation 60
Classical essential gestosis 45
Classification, four stage, ultrasonic appearances 37
Clotting factors 44
CMV lymphoplasmacytic villitis 118
– placentitis 118, 119
Coagulation system 158
Collagen disorder 9
Comparison, different parameters 25
Complement C3 47, 48, 52
– components C3 and C4 44
Concentric sclerosis, vessel wall 102
Congenital adrenal hypoplasia 151
– CMV infection 119
– hepatic fibrosis 11
– infection 117

– leukemia 11
– nephrosis 11
– pneumonia 115
– syphilis 122
– toxoplasmosis 120, 121
Conjugated Δ^5-steroid levels 149
– estrogens 150
Consumptive coagulopathy 8
Cord, accessory lobes 60
–, battledore insertion 60
– blood concentrations 150
– –, hormone levels 149
– plasma levels 149
Crown-rump-length (CRL) 133
Cystic adenomatoid malformations, lung 11
– forms, parasite 102
– kidneys 11
Cytomegalic cells 104
Cytomegalovirus 13, 109, 117
– infection 103
Cytotrophoblast 78
–, increase 66
– proliferation 62

D-AD 135
Damaged syncytial microvilli 66
D-BPD 135
Decidua 30, 31
–, thickening 31
Decidual arteries 58, 69
– –, placental bed 66
– arteriopathy 105
– cells 69, 102
– necrosis 2, 4, 87
– segment, spiral artery 88
– vascular lesions 9
– vessel wall 64
Deciduitis, incidence 115
Decrease, platelets 69
Degenerative changes 6
Dehydroepiandrosterone sulfate (DHA-S) 145
Deposits, complement C3 46
DHA-S 151
Diabetes mellitus 8, 44, 76, 77
– without hypertension 96

Subject Index

Diabetic mother 76
– placenta 8, 78–81
– pregnancy 8
– serum 76
– women 66
Diabetics 77
Diagnostic ultrasound, obstetrics 29
Diaphragmatic hernias 11
Diethylstilbestrol exposure 6
Diffuse chorioangiomatosis 106
– villitis 111
Direct surface anastomoses 10
Discordant disruptive structural defects 10
Disruption complex 14
Disseminated intravascular coagulation 71
DNA content 27
D-weight 135
Dysmaturity 78

Echo free spaces, appearance 37
Echogenic areas, basal plate 37
Echovirus 109
Eclampsia 70, 92, 163
Ectopic gestations 3
Electron micrograph, placental bed spiral artery 91, 95
– microscopy, acute atherosis 94
Endoplasmic reticulum 66, 88
Endothelial cells 69
– –, villous capillaries 80
– lining 92, 94
– –, disruption 92
Endovascular sclerosis, vessel wall 102
Enhancing antibodies 42
Enterovirus 104
Enzyme activity 147
Enzymes 158
EPH gestosis 44, 45
– –, preeclampsia 41
Epiphyseal stippling due to warfarin 6
Erythroblastosis 121
– fetalis 10
Escherichia coli 117
Estetrol concentrations 151
Estradiol excretion 148
Estriol 145

– concentrations 151
– excretion rates 148
– measurements 140
Estrone sulfate 147
Evanescent villi 104
Exogenous influences 5
Experimental toxemia 70
Extensive fibrin deposition 67
– intimal thickening 94
Extrachorial attachment, placental membranes 4
– placentas 110

Factor VIII consumption 69
Fatty change 96
Fetal anomalies 10
– arrhythmias 11
– asphyxia 132, 140
– blood flow measurement 139
– breathing movement examination 139
– capillaries of villi, ultrastructure 66
– death in utero 145
– distress 131
– femur, length 133
– glucose levels 76
– growth retardation 69, 96, 110
– hypersomatism 8
– inhibitory effect 46
– risk 164
– T lymphocytes 46
– vascular circuit 104
– – –, selective involvement 104
– – flow, impairment 11
– visualization techniques 6
– weight 24, 25, 27, 131
– – ratio 27
– – –, toxemia 59
– well-being, assessment 134
Fetomaternal interaction 1
– transfusions 10
Fetoplacental hydrops 10, 11
– –, nonimmunologic causes 10
Fetus papyraceus 10
Fibrin 52, 95, 105
– deposition 37, 92
– deposits 7, 69
– formation 37

Subject Index

Fibrinogen/fibrin 46–48
Fibrinogenesis, activating complement system 51
Fibrinoid 49, 88
– degeneration, maternal decidual arteries 58
– –, media 62
– deposits 43, 46, 51
– formation, first stage 48
– layer 43
– material 88
– necrosis 66
– –, arterial wall 106
–, placenta 46
–, terminal villi 43
–, villi 49
Fibrinolysis 69
Fibrinolytic activity 69, 70
Fibrosis, stroma 102
Fibrotic villi 62
Foamy changes 64
Focal degeneration, trophoblast 7
– disruption, endothelial lining 94
– villitis 121
Folic acid deficiency 7
Free estrogens 150
– fatty acids 76
Functional abnormalities, reproductive tract 1
Funisitis 109, 115

Generalized placental dysmaturity 121
Genetic factors 4
Gestation, monozygous in origin 9
Gestational age 134
– trophoblastic tumours 160
Gestosis 46
Glomerular capillaries 46
Glucocorticoid 145
–, excess doses 145
Glucose 76
Glycogen accumulation 78
– deposits 79
– granules 78
Glycosylation, hemoglobin 77
Golgi apparatus 66
Gorillas 70

Grey scale echography 29
Greyish myometrium 31
Group B β-hemolytic streptococcal infection 122
– – streptococcus 122
– D streptococci 117
Growth hormone 76
– retarded infant 38

Haemolytic-uraemic syndrome 96
hCG levels 162
Hematogenous infection 101, 109, 119
Hemoglobinopathies 111
β-Hemolytic streptococcus 117
Hepatic sinusoids 52
Heroin 6
Herpes simplex 120
– – virus 117, 119
Herpetic infection, newborn 120
Histocompatibility 45
HL-A compatibility 44
– incompatibility 44
HPL screening 131
– test 132
Human placenta 23
– placental lactogen (HPL) 130, 132, 140
– syncytiotrophoblast cell membrane 42
Hybrid placenta 52
– –, nonsensitized hamsters 52
– – – mice 52
– – – rats 52
– –, preimmunized animals 52
Hydatidiform mole 13, 45, 162
Hydramnios 10
Hydropic swelling, Golgi apparatus 66
Hydrops fetalis 45
11 β-Hydroxycorticosteroids 149
16 α-Hydroxylated Δ^4-steroids 149
Hyperbilirubinemia 116
Hyperplacentosis 45
Hypertension 110, 129
–, pregnancy 7
Hypertension/pre-eclampsia 161
Hypertensive placenta(s) 59, 68
– –, ultrastructure 66
– pregnancies 7, 59, 67
– women 66

Subject Index

Hypertrophic endothelial cells 90
Hypoplasia, fetal adrenal glands 145
Hypovascular villi 62, 105
Hysterectomy 86

Ichthyosis 154
Ichthyosis-affected individuals 154
Idiopathic fetal growth retardation 96
– thrombocytopenic purpura 8
IgA 47, 48
IgG 47, 48, 52
IgM 47, 48, 52
Immature villi 78
Immune complex deposition, artery walls 96
– complexes 45
Immunization, fetus in utero 41
Immunofluorescent studies 51, 66
Immunoglobulins 46
Immunologic incompatibility 12
Immunopathological disorders, pregnancy 41
Immunosuppressive agent 158
Implantation 51
– site 51
Incompetent uterine cervix 3
Increased placental residual blood volume 78
– syncytial knots 58
– – sprouts 58
– vascularization 78
Infarction 105, 110
Infarcts 13
Infection 12
Inflammation 4, 12
Insulin 76
–, specific binding 44
Intercotyledonary septa 36
Interlobular septa 34, 35, 37
Intervillous space volume 24
– thrombi 13, 62
– thrombosis 60, 64
Intima, myometrial segments, spiral arteries 94
–, thickening by oedema 95
Intimal cells 94
– –, decidual spiral arteries 69

– –, fat accumulation 94
– thickening 94
Intra-abdominal obstruction, portal circulation 11
Intracapillary erythroblastosis 11, 12
Intraluminal cells 90
– –, ultrastructural features 90
– thrombosis 94
Intrauterine death 10
– deprivation 7
– ectopics 3
– fetal death 3
– growth 153
– – retardation 27, 110, 111, 129
– – retarding 7
– heart failure 10
– – –, abnormal hemoglobins 10
– infection(s) 109, 123, 129
Intrauterine-growth-retarded infant 24
Invasive mole 41
IUGR 129, 131, 136, 138, 140, 161
–, asymmetric 137
–, general screening 132
– screening methods for detection 130
–, symmetric 137

Kernicterus 116
16-Keto-androstenedione 149
Kidneys, patients 96
Klebsiella enterobacter 117

Large intramural cells 88
Leukemia 8
Leukocyte migration inhibition test 45
Linear array scanners 133
Liver, preeclamptic patients 52
Low estrogens 152
Low-birth-weight infants 100, 101
– in term infants 110
Luminal occlusion 102
– wall 69
Lupus erythematosus 96
Lymphangiectasis 11

Macrophages 96
–, amnion 2
–, chorion 2

Macrosomic infant 8
Major estrogen 145
Malignant hypertension 96
Marginal hemorrhages 60
Massive teratomas 11
Maternal anemia 7, 108, 111
– blood, hormone levels 149
– α-fetoprotein (AFP) levels 140
– hypertension 110, 161
– immunological system 45
– lymphocytes 46
– metabolic error 5
– plasma estriol levels 149
– rubella 102
– serum estriol levels 131
– sickle cell trait 8
– systemic disease 7
– urinary estriol excretion 145
– vascular disease 7
– – supply, impairment 7
– venous blood 151
Meconium staining 2
Melanoma 8
Membranes 10, 33
–, early appearance 32
Meningitis 117
Mesangium 46
Metabolic errors 5
Microangiopathy, diabetic placentas 80, 81
Microscopic findings, hypertensive placenta 61
Milder inflammation 12
Mixed lymphocytic culture (MLC) 42, 46
Monitoring 140
Monochorial gestational sac 10
Monochorionic diamniotic membranes 10
Mononuclear cellular infiltrate 93
Morphometric evaluation 17
Multigravidae 45
Multilaminal basement membrane 81
Multiple gestation 9
– pregnancy 161
Mural thrombosis 94
Mycoplasma 123
Myointimal cells 96
Myometrial decidual junction 95

– level 91
– segments 88, 90
– spiral arteries 90

Nasal deformities 6
Necrobiosis, trophoblast cells 102
Necrosis, media 96
Neonatal pneumonia 115
– pneumonitis 118
– sepsis 115, 117
Neuroblastoma 11
Neuroblastomatous metastases, villi 11
New placental proteins, characteristics 159
Nonimmune hydrops 11
Nonimmunologic fetoplacental hydrops 5, 8
Nonsmokers 77
Non-villous migratory trophoblast 87, 90
Normal amniotic fluid DHA 151
– placenta 25
– pregnancy 87
Nuclei, actual content per placenta 24
–, different types 24
Number of capillaries per unit area 23
– of villi per unit area 23
Numeral density 24
Numerous fetal erythroid cell precursors 121

Obstructive lesions, fetal gastrointestinal tract 10
Oedema 95
16α-OH-DHA 148, 151
16α-OH-pregnenolone, excretion 148
16α-OH-DHA-sulfate 149, 150
16α-OH-Δ^5-steroids 152
16α-OH-progesterone 150, 152
17α-OH-progesterone 149
Old hemorrhage 4
Oligohydramnios 3, 10, 14
Oxytocin 153

PAPP-A 157, 163
–, abnormal pregnancy 164
–, normal pregnancy 163
Parenchyma: nonparenchyma, ratio 21
Parenchymal components, placenta 22

Subject Index

- tissue, distribution 21
- weight 24
- – ratio 27
Parental consanguinity 45
Parturition, mode 152
Patas monkey 70
Paternal immunogenetic factor 45
Perinatal death, pathogenesis 61
- mortality 61
Peripheral capillary surface 21, 24
- villous surface 21
- – tissue 22
Perivillous fibrin deposition 64
Perpendicular abdominal diameters (AD) 134
Phenylketonuria 4
Phocomelia, thalidomide 6
Physiologically immature infant 8
Phytohemagglutinin (PHA) 41
Placenta 2, 3, 7, 8, 12, 13, 19, 70, 71, 77, 100, 111, 145
–, amnion nodosum 10
–, appearance 33
–, avillous spaces 35
- circumvallata 1, 6, 61
–, electron-microscopic features 67
- extrachorialis 60
–, granular appearance 32, 33
–, growth retarded infant 26
–, histological features 67
–, infection 2
–, radiograph 36
- with pregnancy toxemia 58
- – vacuolation of trophoblast or stromal cells 5
Placental abnormalities 110
- abruption 162
- activity, arylsulfatase C 147
- anomalies, diabetic pregnancies 9
- barrier, diabetics 77
- bed 86, 87
- – biopsy 63, 86, 88, 89
- – spiral arteries 59, 87, 90–92
- biosynthesis, estriol 146
- circulatory disturbances 105
- development 3, 6
- dysfunction 164

- enzyme activity 146, 147
- findings 6, 11
- – , severe pregnancy toxemia 60
- histopathological changes 121
- histopathology, congenital herpes simplex infection 120
- infarction 58, 60–62, 70, 110
- –, rabbit 70
- insufficiency 110, 129, 137, 140
- labyrinth, giant panda 71
- lesions 10, 70, 101
- localisation 29
- manifestations, fetal distress 2
- –, maternal genetic disease 5
- membrane 17, 26
- morphology 7
- –, diabetics 77
- products, new classification 165
- proteins 157, 159
- septa 86
- shapes 4
- size 52
- sulfate deficiency 145, 146
- sulfatase deficiency 149, 151, 152, 153
- tissue 24, 33
- villi 46, 66
- –, branched tree 51
- –, ultrastructural alterations 66
- villous infarction 2
- weight 27, 131
- –, pure type 59
- –, SGA 21
- – ratio, toxemia 59
Placentas, abnormal configurations 110
–, diabetic women 78
–, gross abnormalities 111
–, smokers 77
Placentitis 13, 113, 117
Plasma cell chorioamnionitis 120
- concentrations, estradiol 149
- constituents 96
Platelet aggregates, trophoblastic surface 46
- aggregation 67
- consumption, local site 51
- deposition 92
Point-counting method 19

Polarized light 49
Polyploidy 5
Potter phenotype 3
PP5 160, 161, 163
–, abnormal pregnancy 161
– levels 162
–, normal pregnancy 161
Preeclampsia 46, 59, 63, 69, 92, 95, 96
Preexisting essential hypertension 59
Pregnant rhesus monkey 70
Pregnenolone-sulfate 147, 149
Premature delivery 162
– rupture, membranes 12
– separation 2, 7
Prenatal diagnosis 151
Preterm infant 2
Progesterone 149
Proliferation, capillaries 81
–, cytotrophoblast 58
–, trophoblastic tissue 41
Proliferative vascular disease 81
– – lesions, placenta 44
– villi 104
Proliferative-necrotic villitis 102, 104
Prostaglandin 153
Protein hormones 158
Protoplasmatic inclusions 103
Pseudomonas aeruginosa 117
Psychomotor retardation 116

Real-time scanners 133
– ultrasound technique 139
Recent hemorrhage 4
Reduced platelet life span 51
Rejected renal transplants 96
– transplants 51
Renal agenesis 3
– allograft rejection 52
– disease 129
– – without hypertension 8
– lesions 46
– transplants 96
Respiratory distress syndrome 2
Retention, dead fetus 3
Reticular deposits, calcium 37
Retrodecidual hematomas 105
Retroplacental hematomas 58, 60, 64

Rhesus incompatibility 10, 45
– monkey 70
Rubella 13, 121
– placentitis 103, 121
– virus 109, 117

Saturated fatty acids 76
Scleroderma 96
Septa 35
Septate uterus 3
Severe inflammation 12
– preeclampsia 95, 163
– toxemia 129
– –, pregnancy 66, 67
Severely hypertensive pregnancies 7
SGA infant 21, 22, 25
– infant's placenta 21, 22, 24–27
Sheep placenta 26
Sicklemia 111
Single umbilical artery 8, 9, 110
Smoking 161
Smoothness, chorionic plate 37
Sorbitol pathway 77
Spiral arteries 69
– –, placental bed 62
– artery 88–90, 94, 95
SP1 157, 158, 160, 161
–, abnormal early pregnancy 159
– – late pregnancy 161
–, early pregnancy 159
–, normal pregnancy 158
Spontaneous early gestational wastage 5
– labor 152, 153
Staphylococcus aureus 117
Stem villous surface 21, 24
– – tissue 22
Stereological analysis 19
Steroid levels, venous cord blood 150
$6\varDelta^5$-Steroid sulfates 150
Streptococcus agalactiae 122
Structural abnormalities, reproductive tract 1
Subchorial thrombosis 110
Subepithelial histiocyte infiltrate 102
Subseptate uterus 3
β-Subunit, human chorionic gonadotrophin (hCG) 159

Subject Index

Sulfatase activity 146
– deficiency 147
Superficial layer, basal decidua 86
Superimposed preeclampsia 59
Suppressor T cells 42
Surface ultrastructure, placentas 67
Syncytial cytoplasm 66
– knot formation, accentuation 7
– knots 105
– rough endoplasmic reticulum, dilatation 66
Syncytiovascular membranes 23, 24, 26
Syncytium, increased loss 58
Syphilis 117, 122
Systemic lupus erythematosus, pregnancy 9

Terminal villus 61
Thalassemia minor 8
Threatened abortion 159
Thrombosis 11
Thrombotic thrombocytopenic purpura 8
Total villous tissue 21
Toxemia 59, 111
–, pregnancy 7, 105
Toxoplasma 13
Toxoplasma gondii 102
Toxoplasmosis 117, 120
Transfer function 26
Transformation rate, peripheral lymphocytes 41
Transplacental carcinogenesis 6
Trophoblast mass 24
Trophoblastic basement membrane 48
– tumours, benign 162
– –, malignant 162
– – –, thickening 58, 62, 66, 81
– invasion, spiral arteries 89
– tumors 160, 162
True infarcts, increased numbers 58
Turner's syndrome 11
Twins pregnancies 45, 134, 164
Twin-to-twin transfusion 9
– – syndrome 11

Ultrasonic appearances 37
– biometric parameters 134

– fetometry 140
– – parameters 133
– placental appearance 29
Ultrasound 130
– fetal biometry 133
– fetometry 136, 138, 139
– –, detection of birth weights 136
– –, efficacy 136
– method 137
– screening 133
Umbilical arteries 78, 110
– artery, absence of one 13
– blood flow, pregnancies with IUGR 139
– cord 106
– –, abnormal insertion 60, 106
– – vasculitis 115
Unfavourable outcome, threatened abortion 159
Unusually long cords 13
– short cords 13
Urinary estrone 148
Urine, hormone levels 148
Uterine leiomyomata 3
– responses 153
Uteroplacental arteries 51
– –, placental bed 86
– –, morphological changes 86
– ischemia 7
– vasculature 110
– vessels 51
– –, walls 51

Vaginal adenosis 6
– carcinoma 6
Validity of a diagnostic test 130
Vascular anastomoses 10
– disease 76
– wall 77
Vasculitis, consistent finding 102
Vasculopathy 9
Velamentous insertion, cord 60
Villi, preeclampsia, features 62
–, surface 48
Villitis 13, 101, 102, 104, 110
–, intrauterine infections 123
Villous abnormalities 110
– dismaturity 106, 111

Villous infarcts 7
– necrosis 58, 120
– pallor 13
– surface densities 23
– –, severe toxemia 68
– tissue, hematogenous contamination 12
Viruses 123

White's classification 77
– group D 77, 78, 80, 81
Widespread subchorial thrombosis 110

X-linked 153

Yellow fluorescence 49